MRE

Materials Research and Engineering
Edited by B. Ilschner and N.J. Grant

Julian Szekely   Olusegun J. Ilegbusi

The Physical and Mathematical Modeling of Tundish Operations

With 124 Figures and 11 Tables

Springer-Verlag
Berlin Heidelberg New York
London Paris Tokyo

Prof. JULIAN SZEKELY
Department of Materials Science and Engineering
Massachusetts Institute of Technology
Cambridge, MA 02139/USA

Dr. OLUSEGUN J. ILEGBUSI
Department of Materials Science and Engineering
Massachusetts Institute of Technology
Cambridge, MA 02139/USA

Series Editors

Prof. BERNHARD ILSCHNER
Laboratoire de Métallurgie Mécanique
Département des Matériaux, Ecole Polytechnique Fedérale
CH-1007 Lausanne/Switzerland

Prof. NICHOLAS J. GRANT
Department of Materials Science and Engineering
Massachusetts Institute of Technology
Cambridge, MA 02139/USA

ISBN-13: 978-1-4613-9628-4   e-ISBN-13: 978-1-4613-9626-0
DOI: 10.1007/ 978-1-4613-9626-0

Library of Congress Cataloging-in-Publication Data
Szekely, Julian, 1934–
  The physical and mathematical modeling of Tundish operations.
  (Materials research and engineering)
  Bibliography: p.
  1. Steel founding—Mathematical models. 2. Engineering models.
I. Ilegbusi, O. J. II. Title. III. Title: Tundish operations.
TS233.S983 1989   672.2   88-29481

Printed on acid-free paper

Typesetting by Asco Trade Typesetting Ltd., Hong Kong

Preface

In recent years it has been recognized that tundishes play a critical role in affecting the quality of the finished steel products. Furthermore, proper tundish design may be even more important in the development of the novel continuous casting processes that are now in varying stages of realization.

Traditionally, physical modeling has played a key role in tundish design, but the recently evolved computational software packages, the readily accessible computational hardware, and, perhaps most important, the growing experience with tackling a broad range of computational fluid flow problems within a metallurgical context have made mathematical modeling an important factor in this field.

Our aim in writing this book has been to bring realistic perspectives to tundish design. The main purpose is to provide a good physical understanding of what is happening in tundishes, together with a realistic discussion of topics that are still not quite clear. The process metallurgist active in this field has many tools at his or her disposal, including mathematical modeling, physical modeling, and measurements on full plant-scale systems. In this monograph we seek to show how these ideas may be combined to provide a good basic understanding and, hence, an attempt at an optimal design.

In writing this book, we must acknowledge our colleagues who are active in this field, and, particularly, the pioneering work of our friend and colleague Professor El-Kaddah, of Professor Yogesh Sahai, and of Professor Alex McLaine. We also acknowledge our colleagues from industry, Dr. A. Vassilicos, Dr. R. Boom, Mr, J. Hlinka, and the Research Group at MEFOS, and many others who have helped us with this endeavor. We also acknowledge Professor D. B. Spalding, who was one of the pioneers of computational fluid mechanics as applied to industrial processes.

The specific tundish designs discussed in the text may well be overridden as new technology (particularly novel casting systems) take hold; however, the methodologies adopted to tackle these problems should be more durable.

Cambridge, MA, July 1988   J. Szekely
                                                                    O. J. Ilegbusi

Contents

Notation

A	Cross-sectional area of jet	L^2
$\mathbf{B}$	Magnetic flux density	(Wb/m^2)
C	Dimensionless concentration (Chapter 3)	
c, C	Concentration	ML^{-3}
C_p	Specific heat at constant pressure	$QM^{-1}T^{-1}$
C_1, C_2, C_μ	Constants in K-ε model of turbulence ($\equiv$ 1.43, 1.92, 0.09, respectively)	
D	Mass diffusivity	L^2t^{-1}
D_{eff}	Effective mass diffusivity	L^2t^{-1}
D_t	Turbulent mass diffusivity	L^2t^{-1}
E	Constant in log law of the wall	
$\mathbf{E}$	Electric field	(V/m)
f	Fractional volume of fluid in computational cell	
F	Dimensionless tracer concentration for pulse injection	
$\mathbf{F}, \mathbf{F}_b$	Body force per unit volume	$ML^{-1}t^{-2}$
F_x, F_y, F_z	Components of $\mathbf{F}$ along x,y,z directions, respectively	$ML^{-1}t^{-2}$
$\mathbf{F}_r$	Froude number	
g	Gravitational acceleration	Lt^{-2}
G_k	Rate of generation of turbulent energy per unit volume	$ML^{-1}t^{-3}$
h	Pool depth	L
H	Height of tundish	L
J	Integral momentum across jet	MLt^{-2}
$\mathbf{J}$	Induced current density	(A/m^2)
k	Thermal conductivity	$QL^{-1}T^{-1}t^{-1}$
k_{eff}	Effective thermal conductivity	$QL^{-1}T^{-1}t^{-1}$
k_t	Turbulent thermal conductivity	$QL^{-1}T^{-1}t^{-1}$
K	Specific turbulent kinetic energy per unit volume	L^2t^{-2}
L	Length scale	L
l_{ex}	Length of exit nozzle (two-dimensional case)	L
l_{in}	Length of inlet nozzle (two-dimensional case)	L
N	Mass flux	$ML^{-2}t^{-1}$
N_p	Rate of particle coalescence	$(No./s)$
N_{pe}	Peclet number	
n_1, n_2	Number density of particles of "size 1," "size 2," respectively	$(No./m^3)$

P	Fluid pressure	$ML^{-1}t^{-2}$
P_0	Reference pressure	$ML^{-1}t^{-2}$
q	Heat flux	$QL^{-2}t^{-1}$
q_v	Volumetric heat generation rate	$QL^{-3}t^{-1}$
q_s	Flux of particles at free surface	$(No./m^2 s)$
Q	Quantity of tracer injected	(Quantity)
r	Radial coordinate measured from center of tube	L
R	Radius of tube	L
R_c	Radius of cylindrical body	L
Re	Reynolds number	
R_p	Radius of spherical inclusion particle	L
$\bar{R}$	Sum of radii of two coalescing particles	L
t	Time	t
t_r	Mean residence time	t
T	Temperature	T
u	Velocity	Lt^{-1}
u',v',w'	Fluctuating velocity components in x,y,z directions	Lt^{-1}
U	Velocity (scale)	Lt^{-1}
U_0	Centerline velocity in jet or velocity of moving plate	Lt^{-1}
U_x, U_y, U_z	Velocity components along x,y,z directions, respectively	Lt^{-1}
$\bar{U}$	Average velocity	Lt^{-1}
U_T	Terminal rising velocity	Lt^{-1}
U^+	Dimensionless velocity	
U^*	Shear velocity	Lt^{-1}
$U_{x,\infty}$	Free stream velocity	Lt^{-1}
$\mathbf{v}$	Velocity vector	Lt^{-1}
V	Velocity (scale)	Lt^{-1}
$\bar{V}$	Volume of fluid in tundish	L^3
V_t	Volume of tundish	L^3
v	Volumetric flow rate	$L^3 t^{-1}$
V_{inlet}	Inlet velocity	Lt^{-1}
V_{max}	Maximum velocity in sectional plane	Lt^{-1}
V_d	Dead volume	L^3
V_m	Perfectly mixed volume	L^3
V_p	Plug flow volume	L^3
v_w	Speed of wave propagation	Lt^{-1}
x	Longitudinal coordinate (distance)	L
x'	Axial distance from inlet; Tundish half-length	
$X3$	Horizontal distance from inlet to exit	L
y	Transverse coordinate (distance)	L
y_p	Perpendicular distance of first grid node from wall	L
y'	Transverse distance from center; Tundish half-width	

y^+	Dimensionless distance from pipe wall	
α	Thermal diffusivity	L^2t^{-1}
δ	Jet width	L
Δ	Displacement thickness	L^2
ε	Rate of dissipation of turbulent energy per unit volume	L^2t^{-3}
κ	von Karman constant in log law of the wall ($\equiv 0.41$)	
μ	Viscosity	$ML^{-1}t^{-1}$
μ_{eff}	Effective viscosity	$ML^{-1}t^{-1}$
μ_l	Laminar viscosity	$ML^{-1}t^{-1}$
μ_L	Liquid viscosity	$ML^{-1}t^{-1}$
μ_t	Turbulent viscosity	$ML^{-1}t^{-1}$
υ	Kinematic viscosity	$ML^{-1}t^{-1}$
Ω	Vorticity	t^{-1}
ρ	Density	ML^{-3}
ρ_f	Fluid density	ML^{-3}
ρ_g	Gas density	ML^{-3}
σ_c	Turbulent Schmidt number	
σ_e	Electrical conductivity	(Ohm/m)
$\sigma_k, \sigma_\varepsilon$	Constants in K-ε model ($\equiv 1.0$, 1.3, respectively)	
σ_s	Surface tension	Mt^{-2}
σ_T	Turbulent Prandtl number	
τ_{yx}	Component of stress tensor: The first subscript denotes the surface on which it acts, and the second denotes the direction of the stress	$ML^{-1}t^{-2}$
τ_w	Shear stress at the wall	$ML^{-1}t^{-2}$
θ	Dimensionless time	

1 Introduction

1.1 Introduction

It is an accepted fact that tundishes play a key role in affecting the performance of continuous casting machines, whether they are of conventional or novel design.

Figures 1.1 to 1.5 show typical single- and multiple-strand tundish designs for conventional casting systems, while Figs. 1.6 and 1.7 depict tundishes used with novel near-net-shape continuous casting technologies.

In conventional continuous casting of slabs or billets, molten steel is teemed from a ladle (say, of 100–300 ton capacity) into a tundish, which is a long trough, say, 3–7 m long, about N1 m wide, and 1 m deep. From this trough the molten steel is fed to the molds of the continuous casting unit.

The nominal residence time in the mold, that is, the mold capacity divided by the metal flow rate, is usually on the order of 3–10 min. However, the "breakthrough time," that is, the minimum amount of time the steel spends in the tundish (e.g., as established by tracer tests), will be less, possibly much less than this ideal value. Indeed, this breakthrough time has been found to depend quite critically on the tundish design.

Recently, there has been growing interest in novel continuous casting systems, which would produce thin slabs, small billets, or even sheet material in one step. These casting machines also have tundishes, which are intermediate vessels between the ladles and the molds. These tundishes tend to be smaller, say, 0.5–2 m long, 0.5–1 m wide, and just a few centimeters or tens of centimeters deep. The corresponding residence time is also less, usually about 0.5–2 min.

Initially, the main function of the tundish has been as a distributor or buffer vessel, which maintains an even, relatively uniform flow as the casting ladle is being emptied and replaced by a new teeming ladle. This is clearly a necessary requirement, because as the casting ladle is being emptied, the metal flow from it will significantly diminish owing to the reduction in head driving the flow.

The tundish evens out this flow in two ways: (1) since the free surface area of the tundish is much larger than that of the ladle, the variations in head will be greatly reduced, and (2) the flow may be more easily regulated from the tundish with a shallower metal depth through the use of a sliding gate or, less satisfactorily, by a stopper rod.

More recently it has been noted that tundishes perform important functions that go beyond being a buffer between the ladle and the mold.

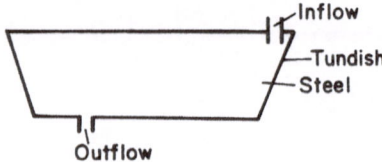

Fig. 1.1. Single-strand tundish design.

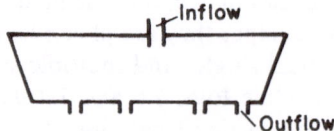

Fig. 1.2. Multiple-strand tundish design.

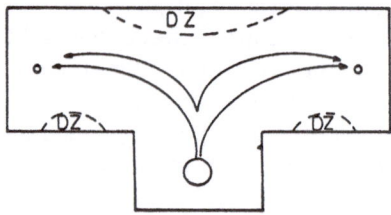

Fig. 1.3. A "T" tundish design after Rehlaender [1].

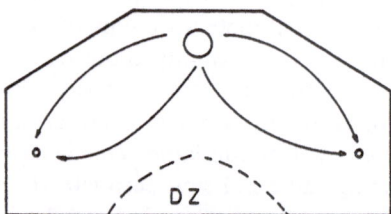

Fig. 1.4. A "DELTA" tundish design after Rehlaender [1].

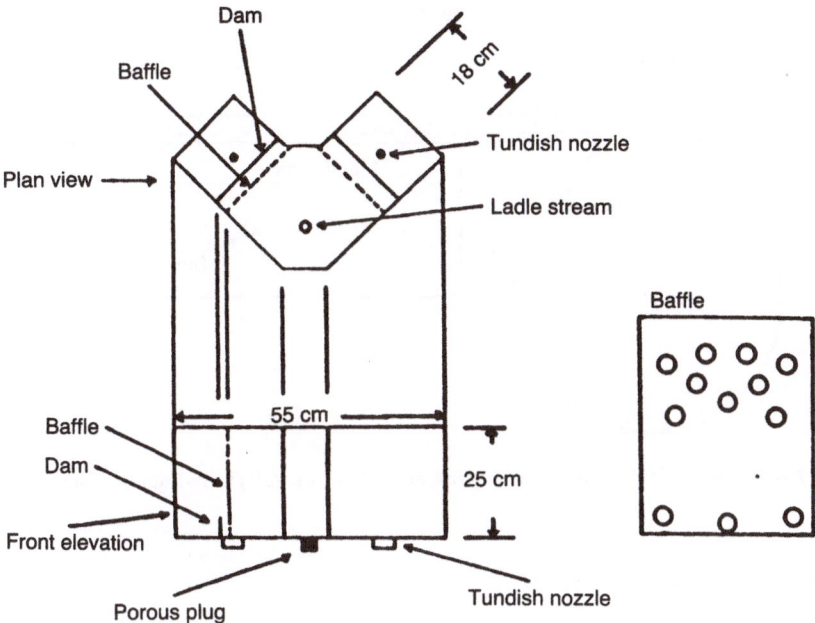

Fig. 1.5. A "V" tundish design after Rehlaender [1].

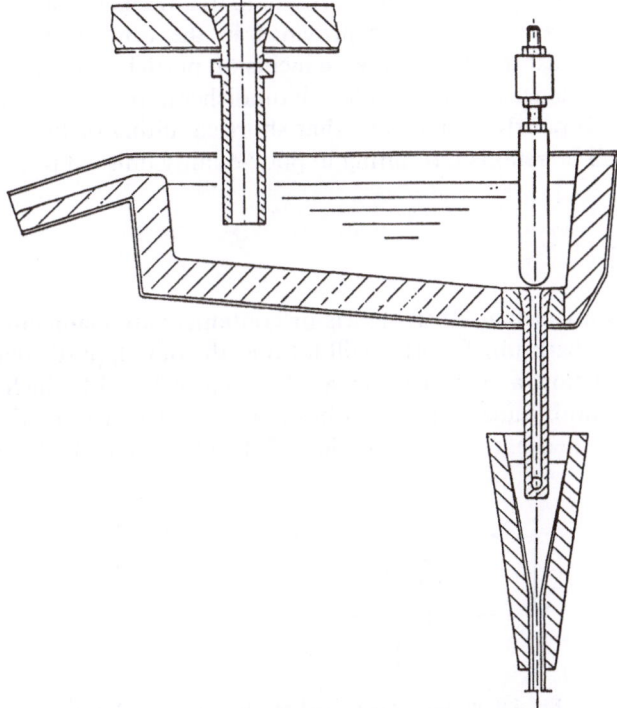

Fig. 1.6. A tundish for the SMS system after Fastert [2].

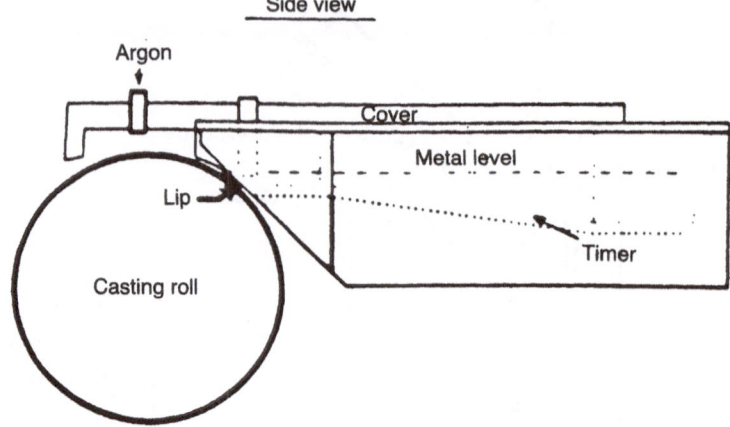

Fig. 1.7. Side view of single-roll caster tundish and roll for the Allegheny-Ludlum system [3].

1.2 Flotation of Inclusions

Perhaps the most important of these additional functions is the flotation of inclusion particles. Even with the best ladle metallurgy practice, some inclusion particles will be retained in the melt; if the ladle metallurgy practice is less rigorous, or non-existent, more inclusion particles will be present in the stream teemed from the ladle. If adequate residence time is being provided in the tundish, there will be an opportunity for at least the larger of these inclusion particles to float out. If, on the other hand, the residence time is short (small tundishes in relation to the metal flow) or if the tundish is poorly designed so that short-circuiting or by-passing occurs, the inclusions will be retained, resulting in poor-quality finished products.

1.3 Vortexing

It is a well-known fact that when vessels or containers are being emptied (e.g., in the emptying of a bathtub), a vortex will form at the outlet, particularly when the liquid level falls below a critical value, as sketched in Fig. 1.8. Such vortexing is undesirable (for both ladles and tundishes), because it can lead to air and slag entrainment, and, hence, to contamination of the cast products. It can be shown

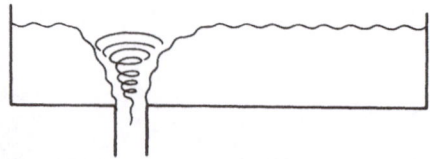

Fig. 1.8. Vortex formation when emptying a tundish.

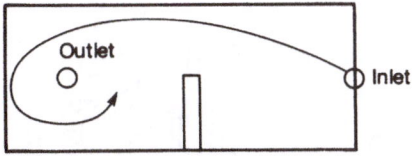

Fig. 1.9. A sketch showing that incorrect baffle location can promote swirl flow and, hence, vortexing.

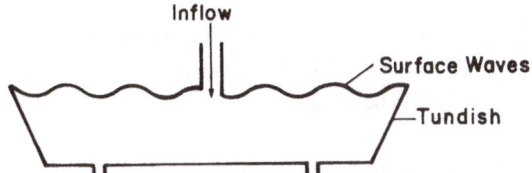

Fig. 1.10. Formation of surface waves due to sudden charge of steel into tundish.

that vortexing, the occurrence of which is inevitable once the liquid level falls below a critical value, may be minimized through correct tundish design and, in particular, through the appropriate placement of weirs and bafflles. By the same token, vortexing may be promoted by the incorrect location of baffles, dams, or weirs, such as sketched in Fig. 1.9.

1.4 Flow Instabilities and Wave Formation

It is an established fact that flow instabilities (i.e., time dependence) may have an adverse effect on the surface quality of continuously cast products. It is also known that when a fluid is suddenly charged into a vessel, or the discharge of a vessel is suddenly initiated, surface waves may form, such as sketched in Fig. 1.10.

Wave formation and flow instabilities may also be controlled or at least influenced by an appropriate tundish design.

1.5 Tundish Metallurgy and Tundish Heating

While at the present time most of the alloying additions tend to be made in the ladle, there are interesting possibilities for making certain additions, such as inclusion modifiers, for example, calcium or calcium–silicon, at the very last minute in the tundish itself. We may also wish to make last-minute temperature adjustments in the tundish, for example, by plasma or induction heating, such as sketched in Figs. 1.11 and 1.12. Such temperature adjustments could become quite critical when we wish to cast with very low superheat.

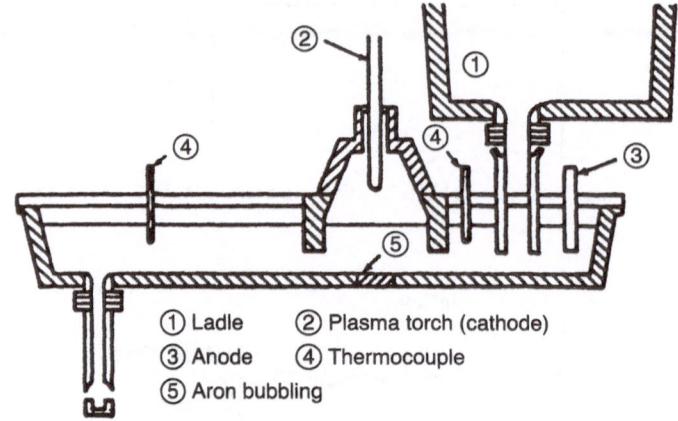

Fig. 1.11. Plasma-heating method in tundish after Saeki et al. [4].

The figure includes the following legend:

① Ladle   ② Plasma torch (cathode)
③ Anode   ④ Thermocouple
⑤ Aron bubbling

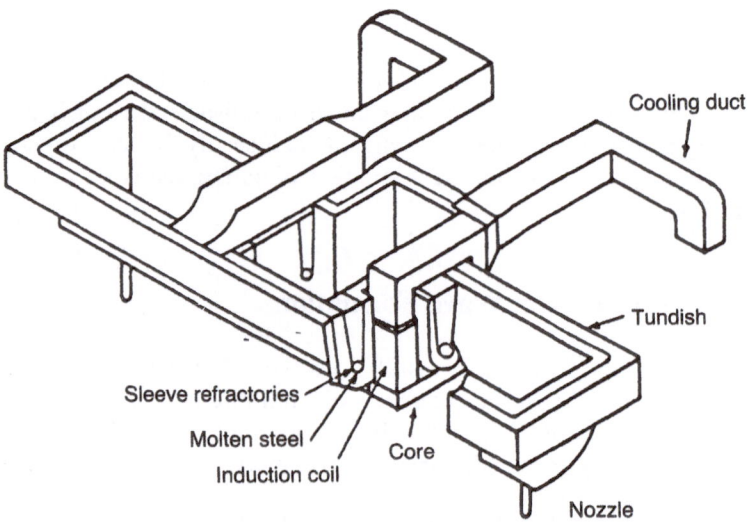

Fig. 1.12. Induction-heating method in tundish after Saeki et al. [4].

The figure includes the following labels: Cooling duct, Tundish, Nozzle, Core, Induction coil, Molten steel, Sleeve refractories.

All these factors point to the need for a more precise tundish design so that these complex objectives can be achieved more easily.

The difference between near-optimal and suboptimal tundish design is clearly illustrated in Fig. 1.13, indicating the marked reduction in magna flux defects in tinplate, which was achieved by an improved tundish design. This figure also shows that most of the defects tend to occur during a ladle change.

Fig 1.14 shows the effect of tundish design on the residence times or "break-through times" for otherwise identical conditions. Here, both real and water model data are shown, and it is readily seen that the use of dams and weirs can markedly

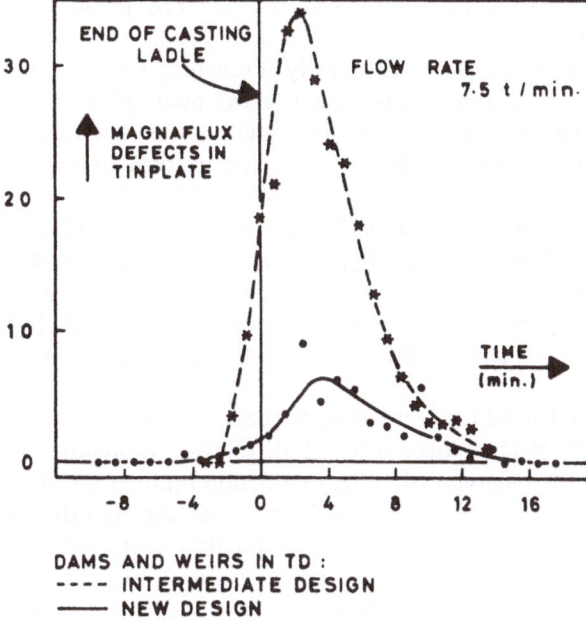

Fig. 1.13. Increase in magnaflux defects at ladle change showing influence of dams and weirs after van der Heiden et al. [5].

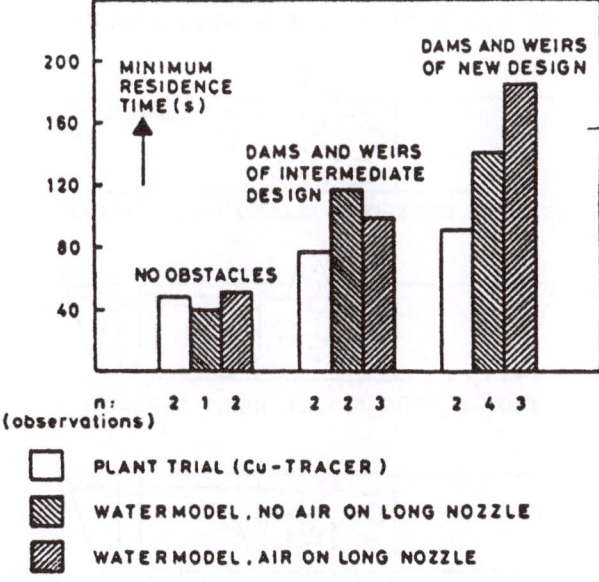

Fig. 1.14. Effect of tundish design on residence times for both real and water models after van der Heiden et al. [5].

increase the minimum residence time and, by inference, promote the flotation of inclusions.

Optimal tundish design is a particularly important topic at the present time, because it is closely related to product quality and metal yield.

This point is readily illustrated by considering the fact that some 200,000–1,500,000 tons of steel would pass through a typical tundish annually, corresponding to finished product values ranging from about $60 to $1,000 million annually (depending on the grades being cast). If we consider that scrap is worth only about $100/ton, then a 1% improvement in yield, that is, a reduction in the rejection rate, would correspond to dollar savings of $0.5–7.5 million per annum. There have been many cases where poorly designed tundishes have caused 10%–30% or higher rejection rates. Under these conditions, the potential savings would become quite spectacular.

In Chapter 2 we discuss the techniques that are available for developing optimal tundish designs. These techniques will include physical modeling, that is, the use of water to simulate the behavior of molten steel, and mathematical modeling.

As a preliminary illustration of what can be done through the intelligent use of mathematical models, Figs. 1.15 to 1.18 show the flow patterns that exist in various tundish configurations. In contrast, Figs. 1.19 and 1.20 show the very detailed velocity maps that one can obtain through mathematical modeling.

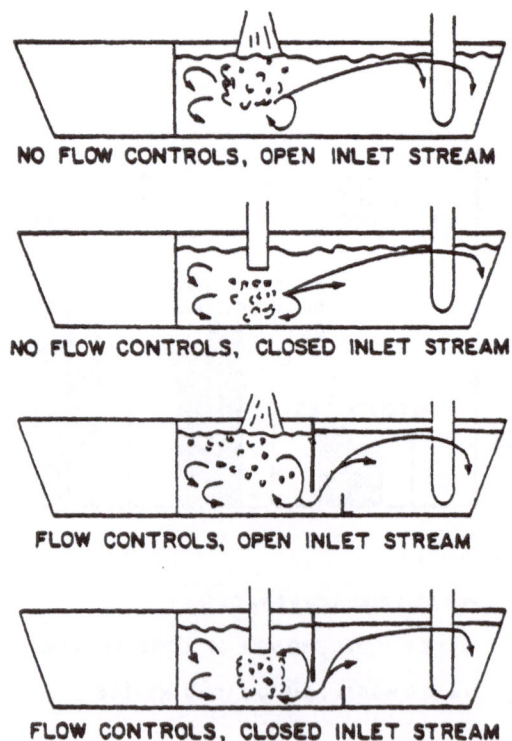

NO FLOW CONTROLS, OPEN INLET STREAM

NO FLOW CONTROLS, CLOSED INLET STREAM

FLOW CONTROLS, OPEN INLET STREAM

FLOW CONTROLS, CLOSED INLET STREAM

Fig. 1.15. Flow patterns in the tundish under various casting conditions after Wilshynsky et al. [6].

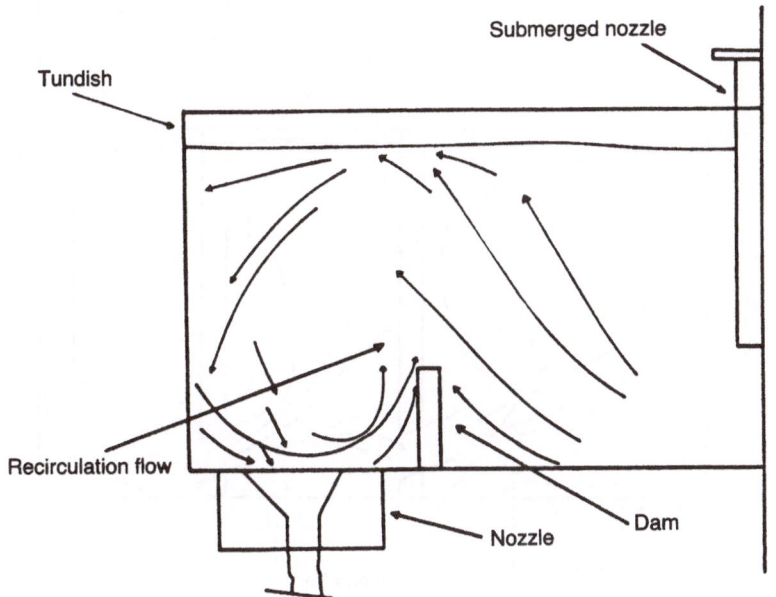

Fig. 1.16. Recirculation effect in a trough tundish with dams [1].

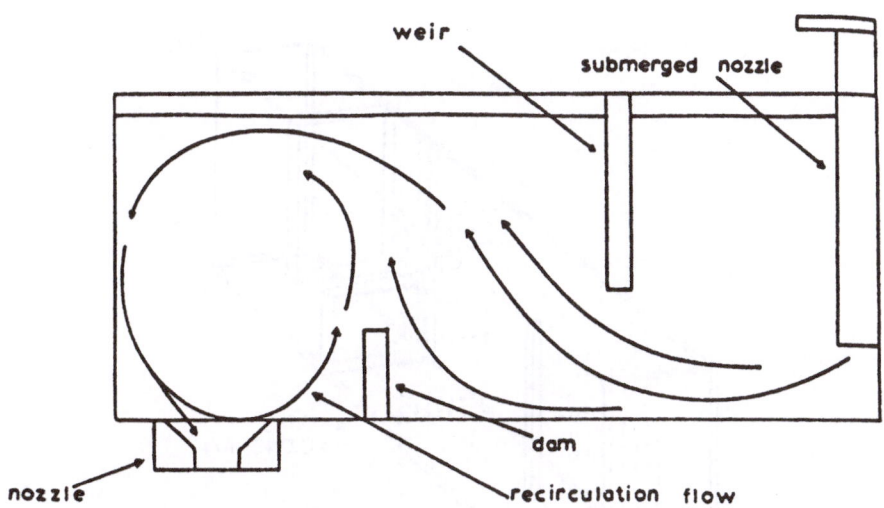

Fig. 1.17. Recirculation effect in a trough tundish with weirs and dams [1].

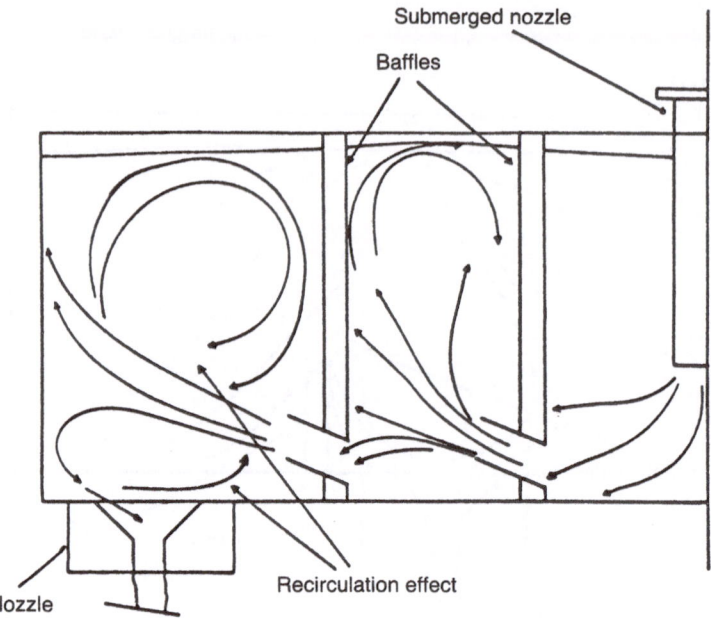

Fig. 1.18. Recirculation effect in a trough tundish with baffles (highly localized mixing design) [1].

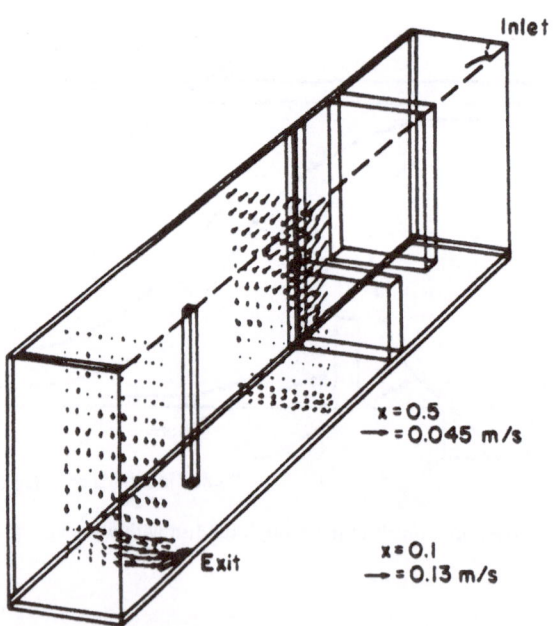

Fig. 1.19. Velocity vectors at two axial planes with one weir and one dam after PCH Physico Chemical Hydrodynamics, 9, J. Szekely et al., "The Mathematical Modelling of Complex Fluid Flow Phenomena in Tundishes," 1987, Pergamon Journals Ltd.

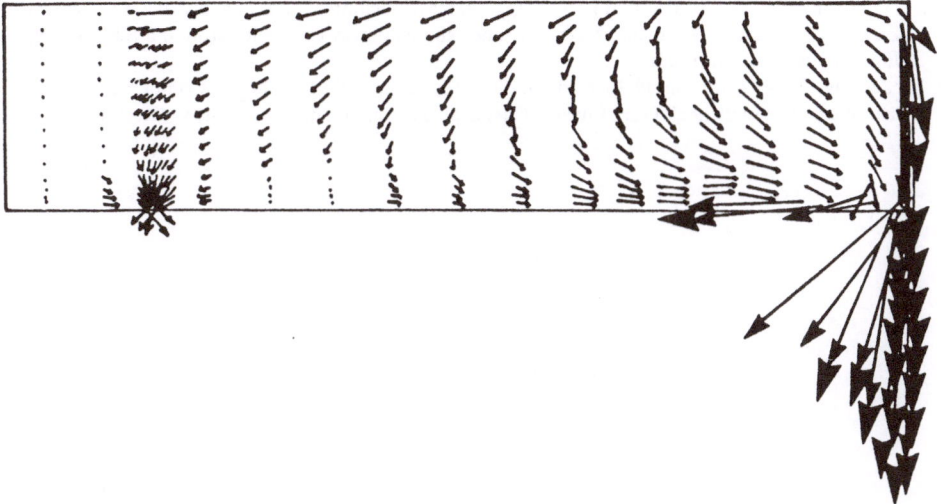

Fig. 1.20. Velocity vectors in a vertical plane near the center of tundish with no flow control ($V_{max} = 2.4$ m/s).

In Chapter 2 we discuss the basic principles that govern the use of these techniques, together with the practical conclusions that may be drawn from them.

It should be stressed that the operation of tundishes represents a complex set of interrelated problems, the solution of which cannot be found in any given discipline, whether it is fluid mechanics or metallurgy or the study of wave phenomena. It is hoped, however, that by using a variety of complementary techniques including water modeling, theoretical fluid mechanics, and plant-scale measurements, we will provide the reader with an improved insight into tundish behavior, and thus aid in the evolution of optimal tundish designs.

References

1. E.M. Rehlaender, "Water Model Studies of Fluid Flow Phenomena and Inclusion Separation in Tundishes," M.S. Thesis, University of Toronto, 1983.
2. H. Fastert, "Latest Developments and Operating Results of the High-Speed Strip-Casting Machine at SMS Schloemann-Siemag, West Germany," Paper No. 219, 89th Annual General meeting of CIM, Toronto, May 1987.
3. J.D. Nauman and D.B. Love, "Controlling the Shape of Cast Stainless Steel Strip," Paper No. 222, 89th annual general meeting of CIM, Toronto (May, 1987).
4. T. Saeki, O. Tsubakihara, A. Kusano, K. Umezawa, and I. Suzuki, "The Roles of Tundishes in Continuous Casting of Steels," Proc. Japan Iron and Steel Society meeting (1987)
5. A. van der Heiden, P.W. van Hasselt, W.A. de Jong, and F. Blaas, "Inclusion Control for Continuously Cast Products," Proc. 5th Int. Iron and Steel Congress, Washington, DC, pp. 755–760 (1986).

6. D.O. Wilshynsky, D.J. Harris, and L. Heaslip, "Water Modelling of Non-metallic Inclusion Separa-
    tion in a Steel Casting Tundish," Notes for tundsish metallurgy course for Iron and Steel Society,
    University of Toronto (1987).

7. J. Szekely, O.J. Ilegbusi, and N. El-Kaddah, "The Mathematical Modelling of Complex Fluid Flow
    Phenomena in Tundishes," PCH Physico Chemical Hydrodynamics, 9, 3–4, 453–472 (1987).

2 Review of Fluid Mechanics Fundamentals

The design of tundishes inherently involves fluid mechanics, because the motion of the melt in the tundish, the behavior of the entrained gases, and the behavior of the inclusion particles are all governed by the laws of fluid flow.

The detailed treatment of fluid flow behavior would be beyond the scope of this text, but it is important to present a brief review of some of the pertinent concepts. For a more detailed study of these phenomena, the reader is referred to specialized texts such as Bird et al. [1], Schlichting [2], and Szekely [3].

2.1 Fluid Flow Regimes: Laminar and Turbulent Flows

When a key dimensionless numbers, the Reynolds number, defined as

$$\text{Re} = \frac{V \cdot L}{v},$$
(2.1)

where

$V \equiv$ velocity scale,

$L \equiv$ length scale,

$v \equiv$ kinematic viscosity

is small (for flow through conduits, the critical value is usually around 2,000), the flow will be laminar, and when Re is large, the flow will be turbulent. There is an intermediate transitional regime between these limiting cases, which is less well defined.

A laminar flow is microscopically time independent, so on plotting the velocity at a particular location as a function of time, we will obtain a horizontal straight line, as sketched in Fig. 2.1. In such situations, transport of heat, mass, and momentum will take place by a molecular mechanism, with the transport coefficients such as the viscosity, thermal conductivity, and diffusivity being determined solely by the chemical composition, temperature, and pressure of the system.

Under laminar flow, the shear stress between adjacent fluid layers, as illustrated in Fig. 2.2, is given by Newton's law, which for a one-dimensional situation is written as

$$\tau_{xy} = -\mu \frac{dU}{dy}.$$
(2.2)

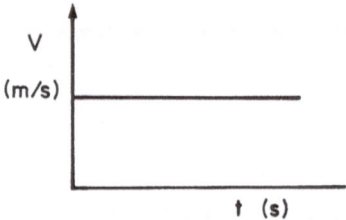

Fig. 2.1. Velocity as a function of time at a particular location in a laminar flow.

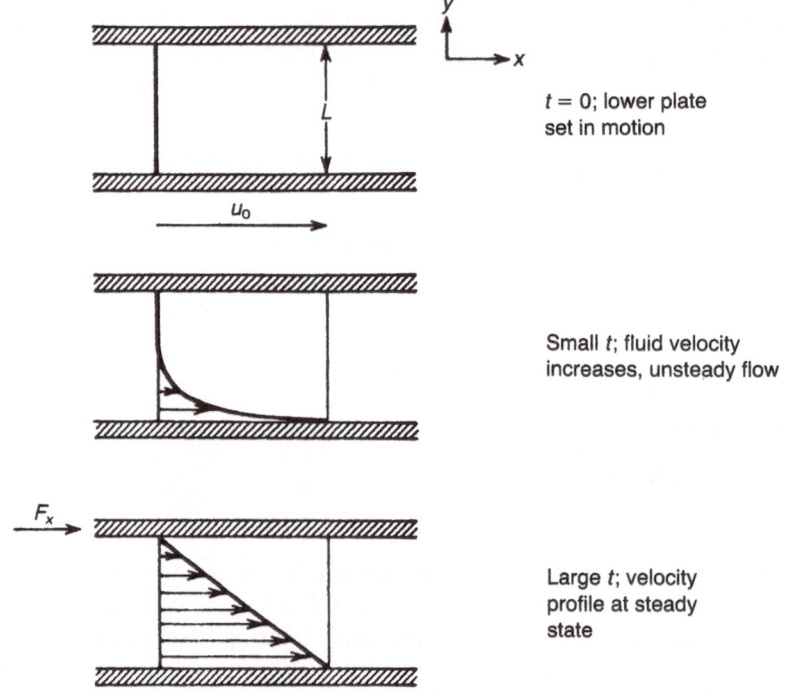

Fig. 2.2. Change of laminar velocity profile with time (upper plate is stationary; lower plate velocity is u_0).

The corresponding relationships between the heat and mass flux and the temperature and concentration gradients may be written as

$$q = -k \frac{dT}{dy} \tag{2.3}$$

and

$$N = -D \frac{dC}{dy}. \tag{2.4}$$

In Eqs. (2.2)–(2.4), μ is the absolute viscosity, k is the thermal conductivity, and D is the mass diffusivity. For multidimensional situations, these relationships are more complex and are available in the previously cited texts.

If we combine the Newtonian definition of the shear stress with a momentum balance, we obtain the celebrated Navier–Stokes equations, which are written in vectorial form in the following manner:

$$\rho \frac{DU}{Dt} = -\nabla P + \nabla \cdot (\mu \nabla U) + \mathbf{F}_b, \qquad (2.5)$$

where

$$\frac{D(\ )}{Dt} = \frac{\partial(\ )}{\partial t} + U \cdot \nabla(\ ) \qquad (2.6)$$

and $\mathbf{F}_b$ is the body force per unit volume.

Equation (2.5) is a complex, multidimensional partial differential equation that relates the velocity and its derivatives to the pressure gradient and to the body force field. In practice, body forces may be produced either by temperature gradients (natural convection) or by electromagnetic forces (electromagnetic stirring or magnetic damping).

Analytical solutions exist for laminar Navier–Stokes equations for special situations; for example, for flow through a circular pipe we have

$$U(r) = -\frac{1}{4\mu}\frac{dP}{dx}(R^2 - r^2), \qquad (2.7)$$

where

$U(r) \equiv$ radially variable velocity,

$\dfrac{dp}{dx} \equiv$ pressure gradient,

$R \equiv$ radius of the tube,

$r \equiv$ radial coordinate measured from center of the tube,

which gives a parabolic velocity profile. However, for the vast majority of cases, even for laminar flow, the Navier–Stokes equations have to be solved numerically, using digital computers. Fortunately, at the present time, there exist several "software packages" for solving these equations. For an experienced fluid mechanician, this would be a routine undertaking for most of the situations that are likely to be encountered.

In contrast to the laminar flow situation, when the Reynolds number is large, the flow becomes turbulent. Turbulence is a manifestation of flow instability. When the flow is turbulent, the velocity at a particular point in the fluid is not steady but will vary, fluctuating with time as illustrated in Fig. 2.3.

We may separate this velocity into a mean ($\bar{U}$) and fluctuating (u') component, and the ratio $u'/\bar{U}$ is regarded as the level of the intensity of turbulence. For many

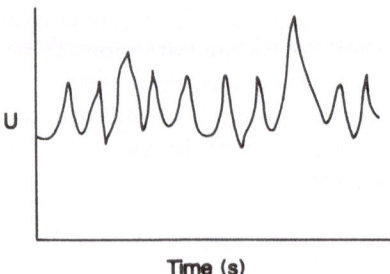

Fig. 2.3. Plot of the fluctuating velocity in turbulent flow.

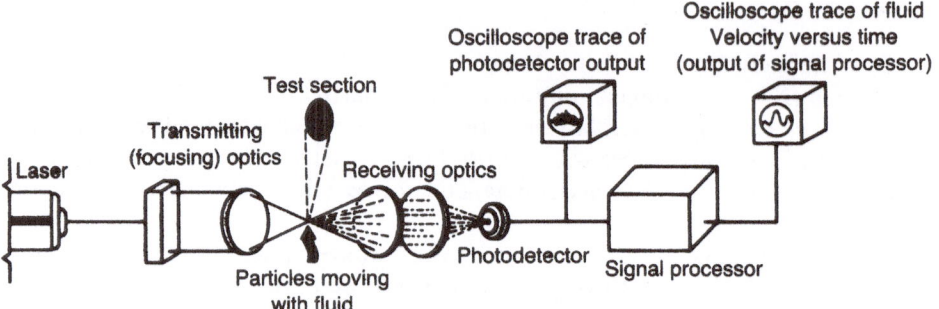

Fig. 2.4. Complete dual-beam laser Doppler velocimetry system.

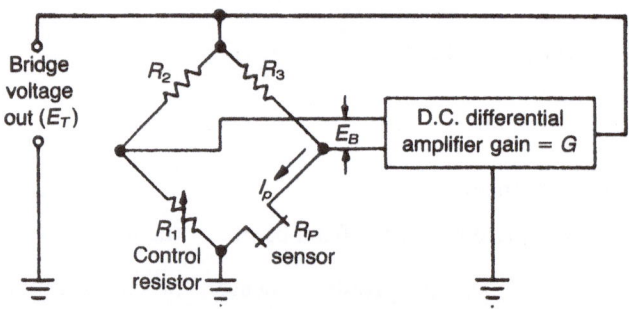

Fig. 2.5. Schematic of constant-temperature hot-wire-anemometer system.

flows, the absolute value of the fluctuating velocity is some 2%–10% of the time-averaged value.

These velocity fluctuations, which can be measured with sophisticated instruments such as a laser velocimeter or a hot-film anemometer, are important manifestations of turbulent flow behavior. Figures 2.4 and 2.5 show sketches of the laser velocimeter and hot-film anemometer, respectively. In many cases, the turbulence is isotropic, meaning that the magnitude of the fluctuations is direction-dependent.

Fig. 2.6. Eddy formation in turbulent flow through a horizontal channel from Szekely [3].

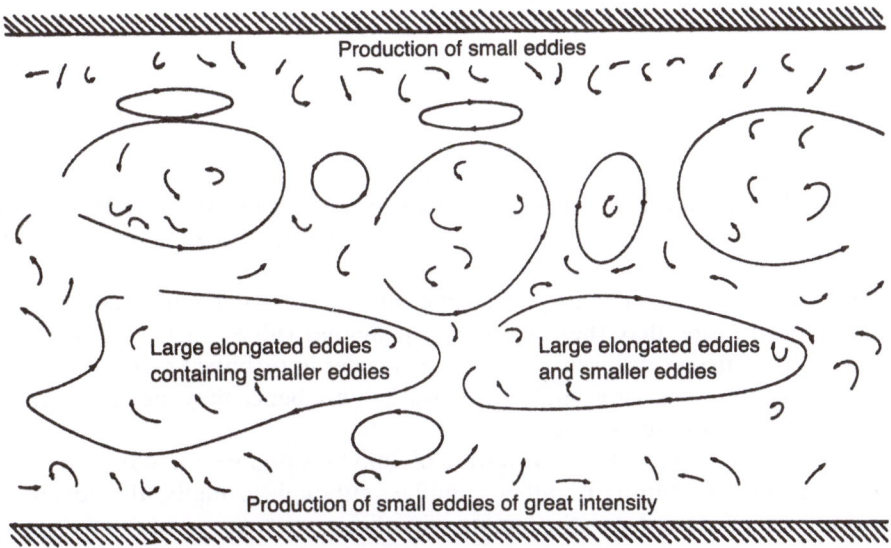

Fig. 2.7. Schematic representation of eddies in pipe flow from Szekely [3].

Another way of representing turbulent flow is to consider the photograph shown in Fig. 2.6 and the corresponding sketch in Fig 2.7. These both indicate the presence of swirls or eddies. Such eddies, which can also be observed when watching the flow of rivers past the abutments of bridges, will form, be destroyed, and then re-form as they travel along with the fluid. The eddies tend to be large in the middle of the vessel, and small in the vicinity of the solid surfaces.

The large eddies, which have dimensions typically some one-tenth to one-third of the vessel characteristic length, are called the macroscale. These eddies contain the turbulent kinetic energy, which is associated with the fluctuating velocity components.

The small eddies, which may have dimensions of typically a hundred microns or so, are called the microscale of turbulence. It is in these eddies that the turbulent kinetic energy is being dissipated. The small eddies may also play an important role in promoting the coalescence of inclusion particles.

A more detailed treatment of the above phenomena could be found in Ref. 4–6.

The very important manifestation of turbulent flow is that the presence of the eddies and the turbulent fluctuations will bring about very effective mixing, because mass (and also heat and momentum) can be readily transported by the eddies from one part of the fluid to the other.

Under turbulent conditions, we can still write formally the following expressions for the shear stress, heat flux, and mass flux:

$$\tau_{xy} = -\mu_{\text{eff}} \frac{dU}{dy}, \tag{2.8}$$

$$q = -k_{\text{eff}} \frac{dT}{dy}, \tag{2.9}$$

$$N = -D_{\text{eff}} \frac{dC}{dy}, \tag{2.10}$$

but here the transport coefficients, that is, the effective viscosity, thermal conductivity, and diffusivity, are no longer molecular properties, but will be a function of the flow system itself. Indeed, each is a sum of the molecular and "turbulent" components.

In general, the turbulent transport coefficients may be some hundred or several thousand times larger than the corresponding laminar values. In a practical sense, this means that dispersion or mixing are much faster in turbulent systems; furthermore, temperature, velocity, and concentration gradients may be rather more difficult to sustain in such systems.

The actual mathematical representation of fluid flow and heat and mass transfer under turbulent conditions is rather complex, although formally, the governing equations are quite similar to the laminar situations.

Thus, we have

$$\rho \frac{DU}{Dt} = -\nabla P + \nabla \cdot (\mu_{\text{eff}} \nabla U) + \mathbf{F}_b, \tag{2.11}$$

$$\rho \frac{DI}{Dt} = k_{\text{eff}} \nabla^2 T, \tag{2.12}$$

and

$$\rho \frac{DC}{Dt} = D_{\text{eff}} \nabla^2 T. \tag{2.13}$$

The principal difference between the two equation systems is that for turbulent conditions the transport coefficients, that is, μ_{eff}, k_{eff}, and D_{eff}, are no longer molecular properties, but will have to be deduced from additional, quite complex relationships, usually termed "turbulence models."

The simplest of these models is to postulate that the turbulent viscosity (and the corresponding thermal conductivity and diffusivity) have a constant value throughout the domain, but one which is 100 or 1000 times larger than the molecular value. Such an approach can give one a good first "ball-park" estimate.

A more sophisticated approach is to use a model such as the ones that estimate both the intensity and the scale of turbulence from transport equations. The most universally applied of such models is the K-ε model, originally proposed by Harlow and Nakayama [7] and later developed by Launder and Spalding [8]. In this model, K is the turbulent kinetic energy of the fluctuating motion, defined as

$$K = \tfrac{1}{2}(\overline{u'^2} + \overline{v'^2} + \overline{w'^2}), \tag{2.14}$$

and ε is the rate of energy dissipation in the turbulent eddies through viscous effects. In this model, the effective viscosity is calculated from the relations:

$$\underset{\substack{\left(\text{effective}\atop\text{viscosity}\right)}}{\mu_{\text{eff}}} \quad = \quad \underset{\substack{\left(\text{molecular}\atop\text{viscosity}\right)}}{\mu_l} \quad + \quad \underset{\substack{\left(\text{eddy}\atop\text{viscosity}\right)}}{\mu_t}, \tag{2.15}$$

where

$$\mu_t = C_\mu \rho K^2 / \varepsilon. \tag{2.16}$$

K and ε are in turn deduced from the following transport equations:

$$\rho \frac{DK}{Dt} = \frac{\mu_t}{\sigma_k} \nabla^2 K + G_k - \rho \varepsilon, \tag{2.17}$$

$$\rho \frac{D\varepsilon}{Dt} = \frac{\mu_t}{\sigma_\varepsilon} \nabla^2 \varepsilon + \frac{\varepsilon}{K}(C_1 G_k - C_2 \rho \varepsilon), \tag{2.18}$$

where G_k is the rate of production of K defined as

$$G_k = \mu_t \left(\frac{\partial U_i}{\partial x_j} + \frac{\partial U_j}{\partial x_i}\right)\frac{\partial U_i}{\partial x_j}, \tag{2.19}$$

and C_μ, C_1, and C_2 are empirical constants.

The conservation equations [(2.17) and (2.18)] can be solved in order to calculate K and ε. Since these equations involve the velocity terms, which are unknown at this stage, this involves a lengthy iterative procedure.

Once the turbulent viscosity is known, the turbulent thermal conductivity and diffusivity may be deduced from

$$k_t = \frac{\mu_t}{\sigma_T}, \tag{2.20}$$

Table 2.1. Selected software
packages for fluid flow
calculations

PHOENICS
FLUENT
TEACH (2D, 3D)
2E/FIX
SOLA/VOF
FIDAP
GENMIX

$$D_t = \frac{\mu_t}{\sigma_c}, \tag{2.21}$$

where σ_T and σ_c are the turbulent Prandtl and Schmidt numbers, respectively.

At the present time, with readily available digital computers and software packages, the solution of turbulent two- and three-dimensional flow problems, such as encountered in tundish operations, is a more or less routine undertaking for an experienced computational fluid mechanician.

Table 2.1 shows a listing of selected software packages suitable for the solution of fluid flow problems. The key problem is the definition of the boundary conditions and the ability to interpret the results in an intelligent manner.

The general nature of the results that one can obtain is sketched in Fig. 2.8 obtained from Deissler [9], where it is seen that in the immediate vicinity of the walls, the flow will be laminar, this region being termed the laminar sublayer. Some distance from the wall we have the core, where the flow is fully turbulent. In the latter region, there is excellent mixing, and it would be difficult to maintain steep gradients in temperature, concentration, or velocity. In between the laminar sublayer and the turbulent core, there exists a buffer or intermediate layer.

In subsequent sections of this text, we shall present an extensive set of computed results describing in detail the velocity, temperature, and concentration fields in various types of tundish designs. For the present, however, let us discuss briefly the implications of these fluid flow fundamentals to tundish design and operation. In the next section, we shall also present some semiquantitative descriptions of certain aspects of tundish behavior, which will contribute to a more complete tundish behavior to be developed subsequently.

2.2 The Implication of Fluid Flow Fundamentals to Tundish Design and Operation

Figure 2.9 shows a schematic sketch of a typical tundish used in conventional continuous casting operations. It is seen that this is a trough some 7 m long, about 1 m deep, and about 1 m wide. Typically, molten steel may be poured through this tundish at a rate of say 60 tons/h, through an inlet nozzle that may be 75 mm in diameter. The two outlet nozzles will also have a diameter of say 75 mm.

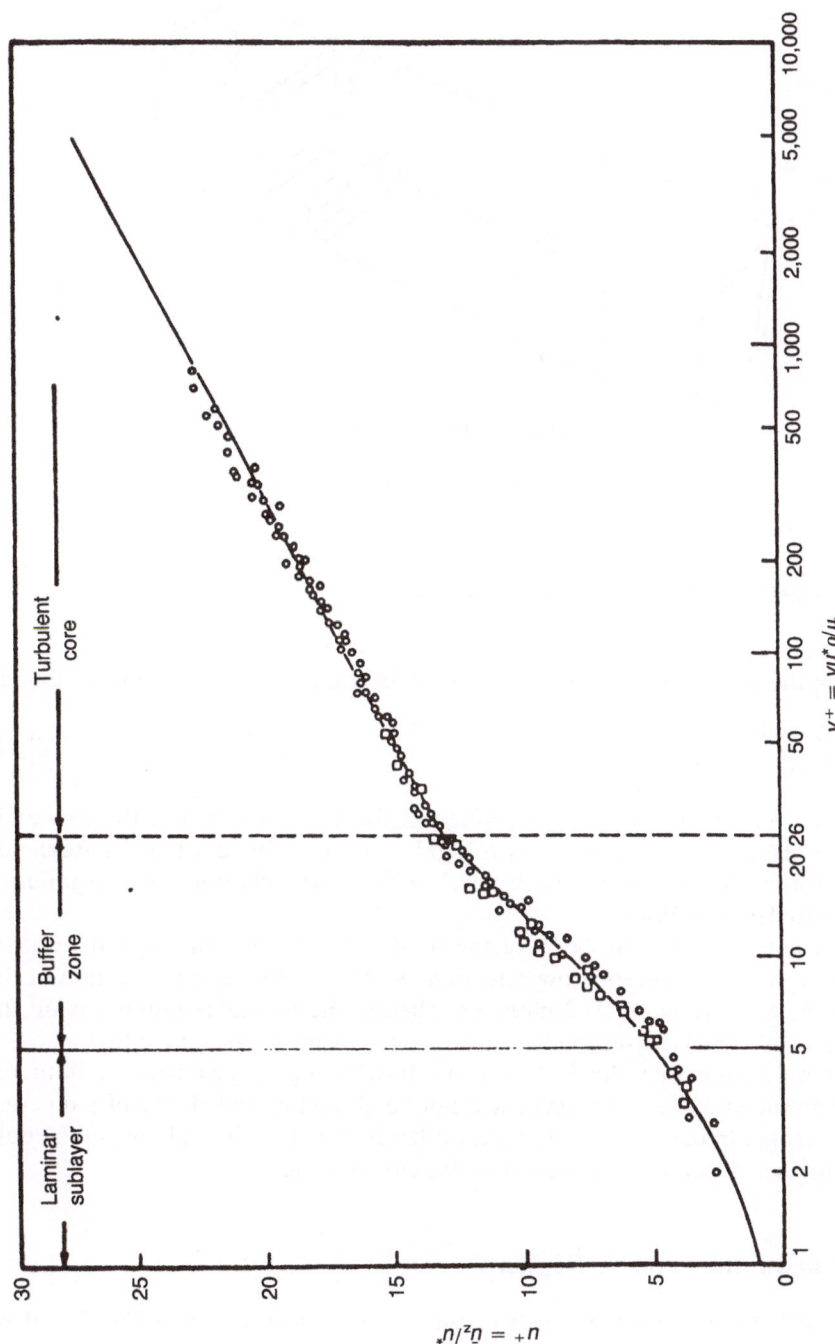

Fig. 2.8. Sketch of the "universal velocity profile" for flow through a pipe, after Deissler [9] from Szekely [3].

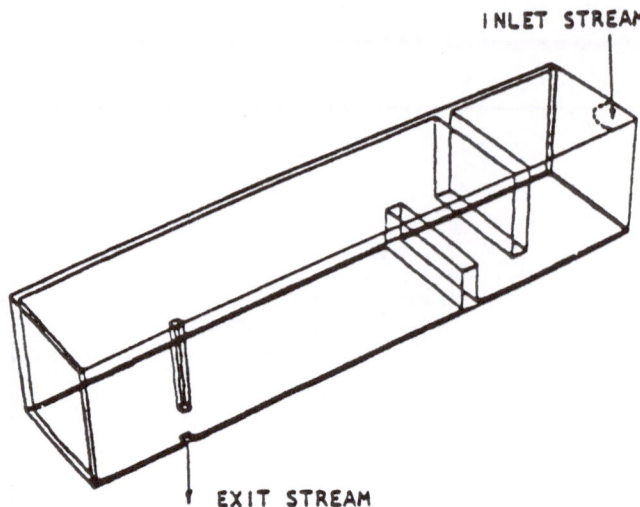

Fig. 2.9. Sketch of a typical tundish system.

2.2.1 Residence Times, Reynolds Numbers

For the conditions considered, the tundish volume is about 7 m^3, and the weight of steel in the tundish may be about 490 tons; hence, the nominal residence time is

$$t_r = \frac{\text{Mass flow rate}}{\text{Tundish mass}} = 8 \text{ min.} \tag{2.22}$$

The linear velocity through the teeming nozzle is 3.2 m/s; hence, the Reynolds number is 2.6×10^5, that is, the flow is highly turbulent. In addition, the Reynolds number for the flow through the exit nozzle is 1.2×10^5, which also corresponds to highly turbulent conditions.

Let us now consider the body of the tundish itself. Here the typical velocity, calculated as the volumetric flow rate, divided by the vertical cross section of the tundish, is in the range of 1–2 cm/s, and, hence, the Reynolds number is on the order of 1,000, which corresponds to transitional or mildly turbulent flow.

We should note that the fact that the flow is highly turbulent in both the feeding nozzle and the exit nozzles will not be altered by the choice of a different tundish design. In contrast, we may expect laminar or transitional flow in the bulk of the tundish, depending on the particular circumstances.

2.2.2 Tundish Inlet Nozzle Region

The teemed stream enters the melt in the tundish, as sketched in Fig. 2.9. If we assume that we are dealing with a "solid" (i.e., unbroken) metal stream, we can consider the following types of idealized behavior:

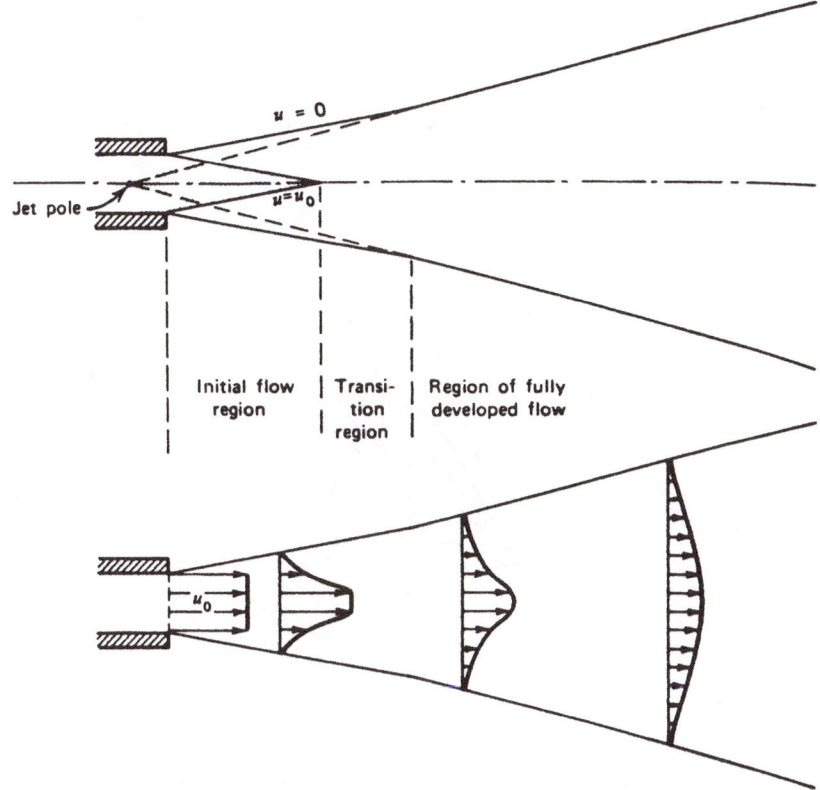

Fig. 2.10. Jet discharging into an infinite fluid.

1) The metal stream may be regarded as a "free jet," i.e., a jet discharging into an infinite fluid, such as sketched in Fig. 2.10. Under these conditions, for a circular jet we have the following relationships:

$$\delta = \text{const} \cdot x, \tag{2.23}$$

$$U_0 = \text{const} \cdot \frac{1}{\delta} \left(\frac{J}{\rho} \right)^{1/2}, \tag{2.24}$$

$$J = \int_0^A \rho U^2 \, dA, \tag{2.25}$$

where δ is the jet width; x is the axial distance; U_0 is the center-line (maximum) velocity; ρ and U are, respectively, the density and velocity of the jet at a distance x from the origin; A is the cross-sectional area of jet at x; and J is the integral of the U component of momentum, which remains constant. The values of the constants expressed in Eqs. (2.23) and (2.24) depend on the jet properties. This behavior is sketched in Figs. 2.11 and 2.12, where it is seen that the incoming fluid entrains the surrounding liquid and is gradually slowed down.

Fig. 2.11. Behavior of a free jet.

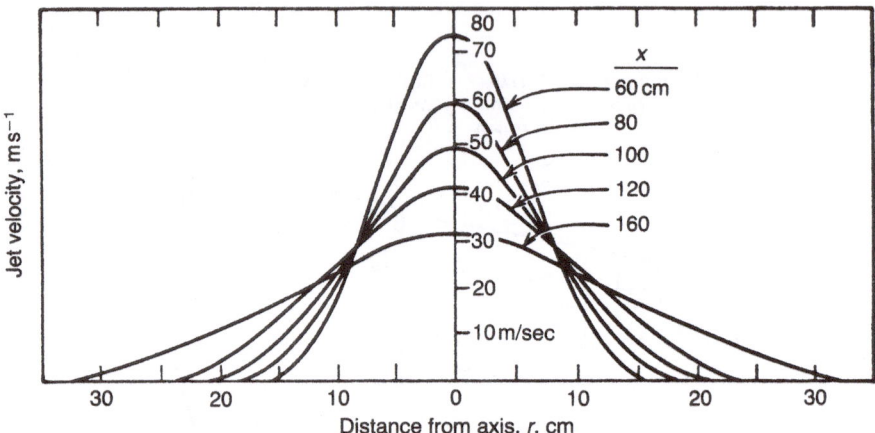

Fig. 2.12. Jet velocity profiles in radial sections at different distances (x) from orifice [3].

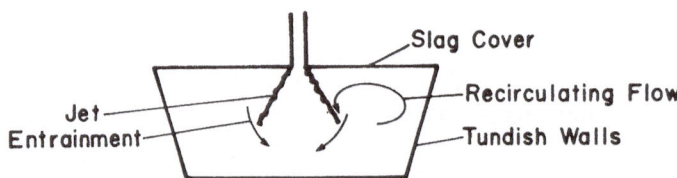

Fig. 2.13. Behavior of confined jet.

2) More realistically, the metal stream may be regarded as a confined jet that is entraining fluid from a confined space. Under these conditions, the resultant flow field will become more complicated, even for the axisymmetric flow situation, resulting in recirculating flow streams, as sketched in Fig. 2.13.

3) If the flow is not axisymmetric, as is the case for most tundishes, the velocity field will be even more complex, with several types of recirculating loops and deflections. In practice, this situation is complicated even more by the entrainment of gases by the teemed stream.

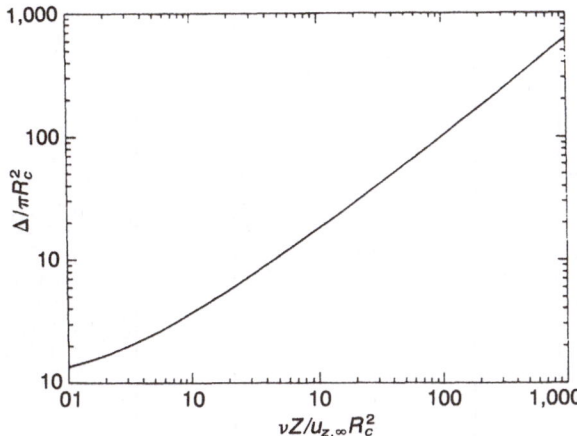

Fig. 2.14. The displacement thickness of the boundary layer formed on a thin cylindrical body from Szekely [3].

2.2.3 Gas Entrainment

When a molten metal stream falls through air or gas, it will entrain some of the surrounding gas with which it is in contact. The extent of entrainment may be estimated with the aid of Fig. 2.14, in which R_c is the radius of the stream, Δ is the cross-sectional area equivalent to the displacement thickness, z is the axial co-ordinate, U_z is the axial velocity, and v is the kinematic viscosity.

These entrained gases will then be released as the stream plunges into the metal body. While the extent of entrainment can be estimated readily, the behavior of the entrained gases in the melt is more difficult to assess, except in qualitative terms. When gases are entrained, the release of the entrained gases will tend to break up the incoming stream and will give rise to a less-well-defined circulation pattern in the vicinity of the pouring stream's entrance.

It should be noted that the rigorous representation of these systems would require the modeling of a multiphase flow situation, where both the bubble size distribution and the bubble trajectories would have to be calculated. While we can adequately describe the behavior of a single bubble in a liquid (particularly very large and very small bubbles), our basic understanding of dispersed two-phase systems is quite limited at present. Indeed, most of the work done to date has involved the development of empirical correlations, primarily for aqueous and organic systems. A great deal more fundamental work needs to be done before these problems can be tackled on a proper scientific basis.

It should be remarked that the behavior of the inlet stream is important not only within the context of a tundish, but that this will take on an even greater significance in the molds of continuous casting systems. Here the nature of the stream can play a major role in affecting the surface quality of the finished products.

Figure 2.15 shows a sketch of the expected flow patterns in molds resulting from various types of tundish operation. It is readily appreciated that the flow pattern

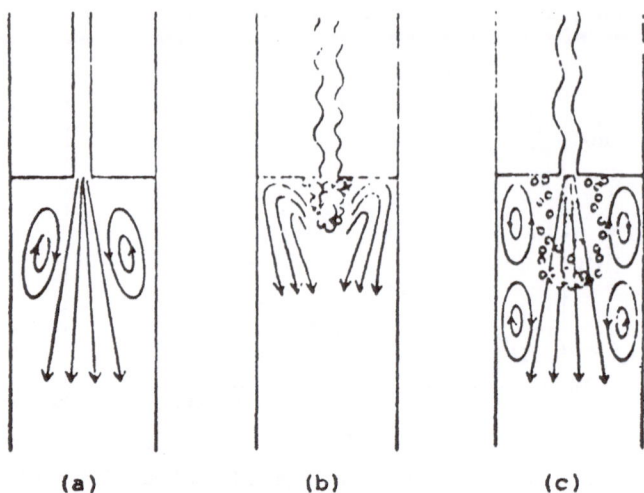

(a)   (b)   (c)

Fig. 2.15. The influence of tundish stream character on the pattern of flow within billet molds (102 × 102 mm or larger in cross section) after McLean [10]: (a) smooth stream, (b) rough stream, (c) ropelike stream.

in the tundish will in turn play an important role in determining the nature of the exit stream, that is, the steam entering the mold.

2.2.4 Vortexing

It is a well-established fact in fluid mechanics that when a tank or container is being emptied, a vortex will form once the liquid level falls below a certain value, such as sketched in Figs. 2.16 and 2.17. This behavior may be aggravated if the flow field in the vicinity of the exit nozzle has a high vorticity, for example, owing to incorrect location of baffes or weirs. Alternatively, the tendency for vortexing may be reduced by the correct placement of baffles or weirs. The definition of the conditions that are conducive to diminishing the tendency for vortexing will be discussed in subsequent chapters.

While the actual onset of vortex formation can be addressed only with the aid of stability analysis, a good indication of whether a vortex will form can be obtained either through the use of physical modeling or through the consideration of the

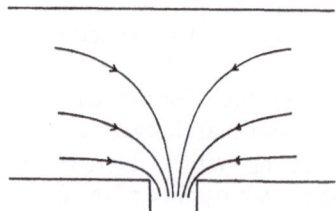

Fig. 2.16. Fluid flow pattern during the draining of a quiescent tank after McLean [10].

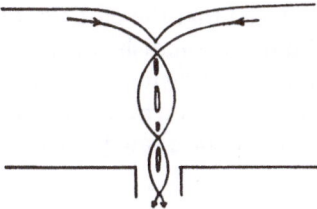

Fig. 2.17. Fluid flow pattern in the presence of a vortex during the draining of a tank after McLean [10].

velocity fields computed using the Navier–Stokes equations. In this latter case, we can define the vorticity

$$\Omega = \nabla \times U, \tag{2.26}$$

in which the component defined in the x-y (lateral plane) is of interest. When the vorticity is high in the vicinity of the taphole, a vortex flow will be established.

One major attraction of mathematical modeling is that the vorticity is easily deduced from the computed flow pattern.

It is an established fact that vortexing would be a highly undesirable behavior because:

1) slag, and possibly gas, could be entrained, causing contamination; and
2) the resulting unsteady pouring stream could cause problems with surface quality of the steel produced.

2.2.5 Inclusion Behavior

One key aspect of tundish operation is that well-designed tundishes will promote the flotation of the inclusion particles. In a *quiescent melt*, U_T, the terminal rising velocity of the inclusion particles, is given by the following relationship:

$$U_T = \frac{2R_p^2(\rho_p - \rho_f)g}{9\mu} \tag{2.27}$$

where R_p is the particle radius, ρ_p is the particle density, ρ_f is the fluid density, μ is the laminar viscosity, and g is the gravitational acceleration.

When the melt is in motion, the rising velocity will have to be added vectorially to both the mean and the fluctuating velocity components.

These quite simple laws have the following consequences.

The typical time of rise of a 20-micron and a 100-micron inclusion particle (of a density about 40% that of steel) from the bottom to the top of the tundish would be about 1900 and 75, respectively. If follows that for a mean residence time of about 4 min, one should be able to float out the larger particles, but that it would be very difficult to float out the smaller ones.

This situation could be improved in the following ways:

1) If one can arrange for an upward-directed flow, which will carry with it the smaller particles.

2) If the inlet stream is so introduced that the particles are not carried down to the bottom, but are introduced into the tundish at a somewhat higher level.

3) If the particles may be agglomerated owing to collisions with each other.

On commenting on this latter point, the rate of particle coalescence depends on the turbulent kinetic energy dissipation rate [11], and is given by

$$N_p = 1.67\bar{R}^3 n_1 n_2 \left(\frac{\varepsilon}{v}\right)^{1/2} \tag{2.28}$$

for two spherical particles, where

$\bar{R} \equiv$ sum of the radii of two coalescing particles,

$n_1 \equiv$ number density of particles of "size 1,"

$n_2 \equiv$ number density of particles of "size 2,"

$\varepsilon \equiv$ rate of energy dissipation,

$v \equiv$ kinematic viscosity.

The turbulent energy dissipation in the pouring box is thought to be comparable to that in a well-agitated ladle system, so it is reasonable to assume that some coalescence will occur in this region. Typical results in agitated ladle systems are shown in Figs. 2.18 and 2.19.

Another important point should be made in this context. While the simple

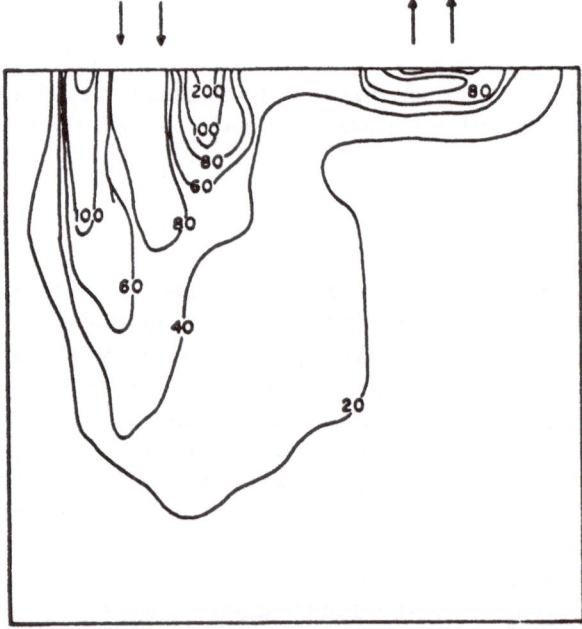

Fig. 2.18. Turbulent energy dissipation (cm^3/s^2) in an R-H vacuum degassing system after Shirabe and Szekely [11].

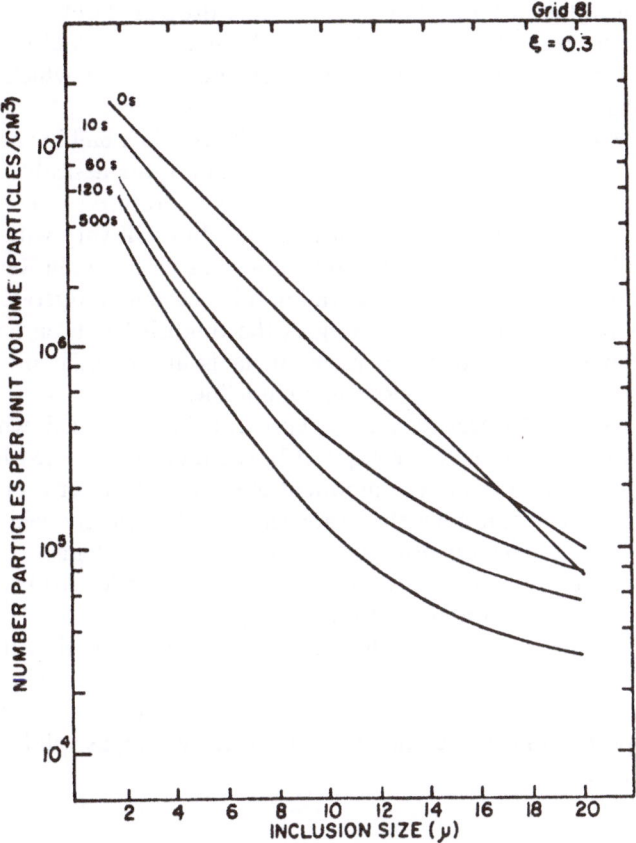

Fig. 2.19. Particle size distribution as a function of time in an R-H vacuum degassing system after Shirabe and Szekely [11].

Stokes-law-type relationship will provide a good indication of the behavior of the inclusion particles, this is far from the complete story. In addition to the problem of agglomeration and coalescence due to turbulence, another key factor has to be considered here also.

The nature of the inlet stream will have a marked influence on the trajectories of the inclusion particles. If the inlet conditions can be so arranged that the inclusion particles are not uniformly dispersed in the pouring box, but are allowed to remain in the vicinity of the top surface (e.g., owing to the presence of gas bubbles or some other arrangements), even smaller particles can be floated out easily, since they will have to travel a much shorter distance in order to reach the top free surface.

2.2.6 Heat Transfer

Heat transfer is an important component of tundish operation. The incoming metal stream will loose heat, primarily by thermal radiation through the slag cover.

Typically, this heat loss will result in a temperature drop of about 10°–20°C during the passage of the melt through the tundish. In the case of multistrand casting situations, the heat loss may be different from strand to strand, which may lead to some operational difficulties.

It may be possible to even out the heat loss by appropriate baffling arrangements; for this reason, a precise knowledge of the flow field is quite desirable.

The additional point has to be made that the temperature of the metal stream entering the tundish fom the ladle is not constant, but will vary with time, as the metal contained in the ladle gradually cools down, as discussed in Ref. [12]. Since in the production of high-quality-steel grades, it is possible to operate with as low a superheat as possible, auxiliary heating of the tundish has been contemplated. The precise understanding of these phenomena is an essential prerequisite for developing rational design and operational guidelines.

The actual heat transfer process may be represented by writing down the convective heat balance equation given in Eq. (2.12). It should be stressed here that the melt velocity distribution is a key component of the heat transfer process, which is markedly affected by it. It follows that knowledge of the velocity field is important in the assessment of the heat transfer behavior of tundish systems.

The boundary conditions usually employed would specify zero heat flux at the tundish walls and thermal radiation from the top slag cover.

If auxiliary heating is performed, this can also be included in the formulation, as will be shown subsequently.

It should be noted that since the flow is quasilaminar in the bulk of the tundish, one should expect significant temperature gradients, which, as will be shown later, is indeed the case.

2.2.7 Mass Transfer

Mass transfer phenomena have to be considered in tundish operation for two reasons:

1) Tracer tests are often used to characterize the flow behavior of tundishes, both as real operating systems and in water models. Tracer dispersion is, of course, a mass transfer process.
2) If alloying additions, or more appropriately, additions to modify inclusions (calcium, calcium–silicon alloys, etc.) are contemplated in the tundish, the dispersion of these is also a mass transfer process.

These processes are governed by the convective mass transfer equation [(2.13)]; whether they take place predominantly by diffusion or by convection is determined by the magnitude of the Peclet number, defined as

$$N_{pe} = \frac{LU}{D_{eff}}, \tag{2.29}$$

where

$L \equiv$ characteristic length,

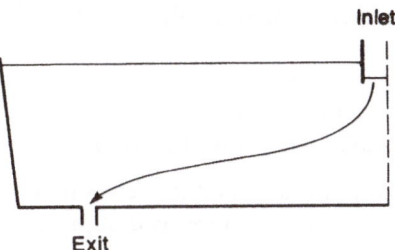

Fig. 2.20. Short circuiting in a tundish system.

$U \equiv$ characteristic velocity,

$D_{eff} \equiv$ effective mass diffusivity.

Since the flow is quasilaminar, the Peclet number will be quite large. For this reason, mass transfer will take place predominantly by convection, rather than by diffusion.

Order-of-magnitude estimates for eddy diffusivity, D_{eff}, show that it may be on the order of $1-50$ cm^2/s in the vicinity of the pouring stream, giving a Peclet number of $0.01-0.1$. This indicates significant diffusive mixing. In contrast, the Peclet number will be very much higher in the remainder of the tundish, signifying that tracer transport there will be due to convection. In this regard, the existence of short-circuiting or by-pass streams, such as those sketched in Fig. 2.20, could have a particularly adverse effect.

We should note that tracer dispersion in the system may be predicted through the solution of Eq. (2.13) for the appropriate boundary conditions. This tracer dispersion will be primarily affected by the velocity field. Indeed, the velocity field predictions may be tested in an indirect manner by comparing the theoretically predicted tracer dispersion rates with those found experimentally.

2.2.8 Unsteady-State Behavior: Wave Formation

Up to the present, we have operated under the tacit impression that tundish operation is basically a steady-state process, with a steady inflow and outflow of molten steel. While this might have been a necessary first approximation, it has to be realized that tundish operation is essentially an unsteady process. When the ladle is being emptied for an initial period, the inflow into the tundish exceeds the outflow, and, hence, the tundish level will rise; conversely, as the liquid level falls in the tundish, the flow rate from the ladle will be exceeded by the outflow, and, hence, the level in the tundish will fall. Indeed, one of the key functions of the tundish is to even out these flow variations.

However, the key point that must be made here is that this essentially transient mode of tundish operations has the following two consequences:

1) There may be significant variations in the melt velocities about the mean, and, in particular, as the tundish level is being raised, the flow pattern may differ quite

significantly from the converse situation, that is, when the tundish level is being lowered.

2) Perhaps a more important consequence of this transient operation is that whenever there is a change in the input and output conditions, there will be an instability, and surface waves will form, such as sketched in Fig. 1.10.

These waves propagate very rapidly, and the resultant disturbances may have an adverse effect on the quality of the continuously cast products. This is an important, not fully explored aspect of tundish operations.

2.3 Concluding Remarks

In this chapter, we have sought to introduce the reader to some of the fundamentals of fluid mechanics and transport phenomena, as applied to tundish design and operation. Owing to space limitations, this introduction has to be somewhat cursory; nonetheless, it is hoped that this will give the interested reader a "flavor" of the fundamentals, which will be used in subsequent sections of the text to explore tundish behavior in greater detail.

2.4 Summation

At this point we ought to summarize what we have learned regarding fluid flow fundamentals.

1) We defined the Reynolds number as

$$\frac{\text{Characteristic length} \times \text{characteristic velocity}}{\text{Kinematic viscosity}}$$

and have shown that this number is very high, in both the inlet and the exit regions, giving rise to highly turbulent conditions there. However, the Reynolds number will be quite small in the bulk of the tundish, producing laminar or transitional flow. Where flow is turbulent, we have excellent mixing, but mixing will be poor in the laminar regions.

2) We developed the Navier–Stokes equations, which in conjuction with turbulence models provide the basis for computing the velocity fields in tundishes. We should note here also that the Navier–Stokes equations provide the fundamental justification for the use of physical modeling approaches.

3) We stated the basic laws of heat transfer and mass transfer, which enable us to calculate these processes in tundishes.

4) We examined a number of particular flow situations which are relevant to tundish operation, including plugging streams, air entrainment, confined and free jets, vortexing and recirculation, the factors that govern the flotation of inclusions, and wave formation inherent in the unsteady operation of tundishes.

The material described in this chapter, together with the techniques to be described in Chapter 3, provide the fundamental tools for studying tundish behavior.

References

1. R. Bird, W.E. Stewart, and E.N. Lightfoot, Transport Phenomena, Wiley, New York (1960).
2. H. Schlichting, Boundary Layer Theory, Pergamon Press, London (1955).
3. J. Szekely, Fluid Flow Phenomena in Metals Processing, Academic Press, New York (1979).
4. J.O. Hinze, Turbulence, McGraw-Hill, New York (1959).
5. G.K. Batchelor, The Theory of Homogeneous Turbulence, Cambridge University Press, New York (1963).
6. J. Lumley and H. Tennekes, A First Course in Turbulence, MIT Press, Cambridge, MA (1972).
7. F.H. Harlow and P.I. Nakayama, "Turbulent Transport Equations," Phys. Fluids, 10, 2323 (1967).
8. B.E. Launder and D.B. Spalding, "The Numerical Computation of Turbulent Flows," Comp. Methods in Applied Mech. and Engg., 3, 269 (1974).
9. R.G. Deissler, NASA Tech. Note D-2138 (1950); NACA Rep. 1210 (1955).
10. A. Mclean, Course Notes on Tundish Metallurgy for Iron and Steel Society, Toronto (1987).
11. K. Shirabe and J. Szekely, "A Mathematical Model f Fluid Flow and Inclusion Coalescence in the R-H Vacuum Degassing System," Trans. ISIJ, 23, 465 (1983).
12. O.J. Ilegbusi and J. Szekely, "Melt Stratification in Ladles," Trans. ISIJ, 27, 563 (1987).

3 The Physical Modeling of Tundish Systems

In Chapter 2 we discussed the fluid flow fundamentals pertaining to tundish operation. At this point, we may proceed in two possible ways:

1) Solve the differential equations numerically, using a digital computer.
2) Develop a physical model (e.g., a water model) of the process.

These two approaches are complementary. At the present time, it is not practical to model everything mathematically, although computer modeling is becoming more attractive with the more ready availability of relatively inexpensive computers and software packages. Furthermore, the data that may be derived from mathematical modeling work can be put to much more effective use through careful interpretation of the results.

In this chapter, we shall address the principles and practice of physical modeling, and we shall also cover the main concepts of continuous flow systems and tracer results.

3.1 The Principles of Physical Modeling

The objective of physical modeling is to represent a system to be modeled by changing the materials to be handled, and also the scale of the operation, so as to achieve realistic representation, but at the same time to allow the measurements to be made more conveniently and in a cost-effective manner. In this way, tundish systems or, more specifically, certain aspects of tundish operations may be modeled using water and plexiglass, and possibly buoyant solid particles.

The principles of physical modeling may be simply derived from the differential equations that govern fluid flow and the associated heat transfer and mass transfer phenomena.

Let us consider the isothermal flow of a fluid through a conduit of arbitrary shape. This process may be described in terms of the Navier–Stokes equations, which, as discussed in the preceding chapter, may be written in the following form:

Equation of Continuity:

$$\nabla \cdot v = 0; \tag{3.1}$$

Equation of Motion:

$$\rho \frac{Dv}{Dt} = -\nabla P + \mu \nabla^2 v + \rho g. \tag{3.2}$$

We shall define some dimensionless variables thus:

$$v^* = \frac{v}{V}; \quad p^* = \frac{P - P_0}{\rho V^2}; \quad t^* = \frac{tV}{L};$$

$$x^* = \frac{x}{L}; \quad y^* = \frac{y}{L}; \quad z^* = \frac{z}{L};$$

$$\nabla^* = L\nabla = \left(\delta_1 \frac{\partial^2}{\partial x^*} + \delta_2 \frac{\partial}{\partial y^*} + \delta_3 \frac{\partial}{\partial z^*} \right);$$

$$\nabla^{*2} = L^2\nabla^2 = \frac{\partial^2}{\partial x^{*2}} + \frac{\partial^2}{\partial y^{*2}} + \frac{\partial^2}{\partial z^{*2}};$$

$$\frac{D}{Dt^*} = \left(\frac{L}{V} \right) \frac{D}{Dt};$$

where P_0 is some convenient reference pressure, t is the characteristic time, L is the characteristic length, and V is the characteristic velocity.

We may rewrite Eqs. (3.1) and (3.2) in terms of the preceding dimensionless variables by setting $v = v^* V$, $P - P_0 = P^* \rho V^2$, etc.:

$$\frac{1}{L} \nabla^* \cdot v^* V = 0, \tag{3.3}$$

$$\rho \left(\frac{V}{L} \right) \frac{D}{Dt^*} (v^* V) = -\left(\frac{1}{L} \nabla^* \cdot P^* \rho V^2 \right) + \mu \frac{1}{L^2} \nabla^{*2} (v^* V) + \rho g. \tag{3.4}$$

Multiplication of Eq. (3.3) by L/V and Eq. (3.4) by $L/\rho V^2$ gives

$$\nabla^* \cdot v^* = 0, \tag{3.5}$$

$$\frac{Dv^*}{Dt^*} = -\nabla^* P^* + \left(\frac{\mu}{LV\rho} \right) \nabla^{*2} v^* + \left(\frac{gL}{V^2} \right) \frac{g}{g}. \tag{3.6}$$

Note that in these dimensionless forms of the equations, the "scale factors," that is, those variables describing the overall size and speed of the system and its physical properties, are often concentrated in two dimensionless groups. These groups occur so often in engineering studies that they have been given names in honor of two of the pioneers in fluid mechanics:

$$Re = \left(\frac{LV\rho}{\mu} \right) = \text{Reynolds number}, \tag{3.7}$$

$$Fr = \left(\frac{V^2}{gL} \right) = \text{Froude number}. \tag{3.8}$$

If the two systems (the prototype and the model) are geometrically similar, that is, if they have the same shape (more precisely, if all the key dimensions are in the same ratio), then the same equations will apply and the only difference will be introduced through the scaling factors, that is, the Reynolds number and the Froude number, which we have just defined. In other words, if the two systems are geo-

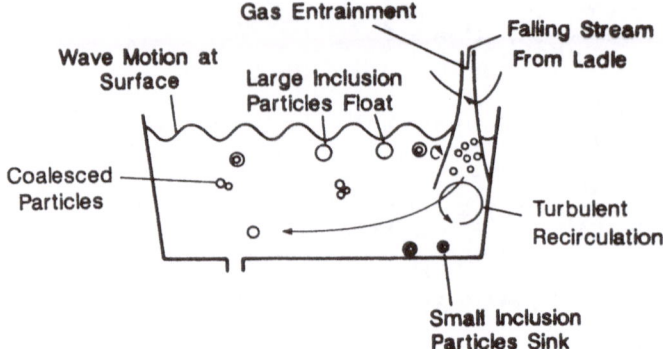

Fig. 3.1. Some processes occurring in a typical tundish system.

metrically similar, they will behave identically, provided the Reynolds and the Froude numbers are the same.

It follows that by the choice of the nature of the model fluid and linear dimensions for the equipment, one can, in principle, achieve a realistic representation of the flow situation.

These arguements pertain to a single-phase flow system, in the absence of free surfaces.

In reality, tundish systems can be rather more complex. As sketched in Fig. 3.1, we can see that

1) The entering metal stream will entrain gases, so that we have to deal with the evolution of gas bubbles.
2) The behavior of inclusion particles, which will in part flow; in part participate in a process of coalescence; in part sink; and in part just be carried along by the metal stream.
3) We may have to deal with wave motion at the top free surface of the melt.

In order to represent these phenomena, additional and more complex differential equations and boundary conditions will have to be invoked, which in turn gives rise to some additional dimensionless groups. A partial list of these groups relevant to tundish modeling is given in Table 3.1.

The important practical consequence of this rather large number of groups is that, whereas the simple single-phase situations are quite easily modeled, the representation of the more complex, multiphase systems in terms of physical modeling would be quite difficult. This is the case, in particular, for systems involving surface phenomena (viz., the Weber and Morton numbers), because the surface tension of liquid metals is very different from that of water.

Indeed, Table 3.2, taken in part from course notes by McLean [1], shows some of these key dimensionless numbers for steel and water in identically sized tundishes.

The important point that emerges from this table is that if the water model is of the same size (and of course the same shape) as the prototype, simple one-phase situations can be quite easily simulated (the Froude and the Reynolds numbers are

Table 3.1. Some dimensionless groups of relevance to tundish modeling

Group	Definition	Application
Reynolds	$\rho L U/\mu$	Fluid flow
Froude	U^2/gL	Wave and surface behavior; pouring stream
Modified Froude	$\rho_g U^2/[(\rho_L - \rho_g)gL]$	Behavior of gas/liquid system
Peclet	$LU/\alpha, LU/D$	Forced convection; mass transfer
Weber	$\rho L U^2/\sigma_s$	Bubble formation; atomization of liquid jets
Morton	$g\mu L^4/\rho L \sigma_s^3$	Velocity of bubbles in liquids
Schmidt	$\mu/\rho D$	Mass transfer
Prandtl	$\mu C_p/k$	Forced and free convection

Table 3.2. Calculated values for various dimensionless groups in the steel and and water model systems (from McLean [1])[a]

Number	Steel system	Full–scale water model
Reynolds	1	1.1
Froude	1	1.0
Weber	1	3.1
Morton	1	44.0
Modified Froude	1	7.0

[a] Values are normalized with respect to the steel system.

almost identical for steel and a full-scale water model). However, aspects of the problem, such as gas-bubble evolution, the coalescence of inclusions, slag entrainment, and possible slag–metal mixing would be very difficult, if not impossible, to simulate using water models.

Furthermore, owing to the disparity in the Prandtl numbers, heat transfer problems, including circulation due to thermal natural convection, would also be difficult to model physically.

It follows that simple streams that are not affected by the entrained gas streams are quite readily modeled physically; indeed physical modeling could be a very cost-effective way of selecting optimal dam, gate, or weir arrangement for these systems.

However, if we wish to model other aspects of the tundish flow problem, such as

the coalescence and flotation of inclusions;
gas entrainment, gas evolution, and evolved gas bubbles' effect on the flow field;
any form of heat transfer, including various forms of auxiliary heating; or
mass transfer, particularly the rate of dissolution and dispersion of alloying elements;

then the use of physical models becomes more problematic. Physical models may still be used, but the results must be interpreted with caution.

We should reiterate here the principles of physical modeling, and the practical consequences of these to tundish problems.

A physical model and the prototype (the real system) are rigorously similar, if we meet the criteria for

geometric similarity (the model and the prototype are of identical shape, but not necessarily on the same scale); and

dynamic similarity (the forces acting on the system are of the same ratio). In practice this can be attained if the key dimensionless numbers are the same.

What does this concept of similarity mean in tundish practice?

1) We can model the bulk flow effects by building a full-scale plexiglass tundish model and use water as the working fluid. This would satisfy the Froude number equality requirement exactly, and the Reynolds numbers close enough. Such a model should also give us a good idea of the large gravity waves (no surface tension effects) and also of vortex formation. Within this context, we can also model the dispersion of tracers; this being an important point, because the use of tracers is an excellent tool in process analysis.
2) The modeling of air entrainment, and particularly the evolution of the entrained gas bubbles, will be qualitative at best, but should give one an idea of the behavior of the system.
3) By the same token, the modeling of inclusion behavior will be qualitative, since the physical modeling approach will not be able to represent the various phenomena that are affected by surface tension, such as coalescence or incorporation into the top slag.
4) The modeling of heat transfer phenomena, such as heat losses and the role of auxiliary heating, will be difficult if not impossible, because of the great disparity in the values of the Prandtl numbers.
5) Finally, the modeling of more "exotic" tundish configurations, for example, involving magnetic damping and induction heating, would be impossible using water due to the differences in electrical conductivities.

With these caveats, physical modeling has proven itself to be a very effective tool in optimizing tundish design and operations, providing very useful insights. Some key results from physical modeling will be discussed in Chapter 4.

3.2 Continuous Flow Systems and Tracer Studies

When a fluid flows through a vessel or a container in which it undergoes a chemical change, it is important to establish the time spent in the system by individual fluid elements. The "average time," or nominal holding time, of the fluid in the system is easily calculated from the definition

$$t_r = \frac{\text{Volume of fluid in vessel}}{\text{Volumetric rate of fluid flow}}.$$

However, it is frequently found that some individual fluid elements may spend a longer, and others a shorter, period of time in the system. This departure of actual residence times from the mean, that is, the spread or distribution of residence times,

is an important characteristic of the system and influences appreciably its performance as a reactor or contacting surface.

The residence time distribution of a fluid flowing through a vessel can be determined by means of tracer techniques. Basically, these involve the addition of a tracer
(e.g., dye, radioactive material, or chemical substance) to the stream entering the
vessel, and then measurement of the concentration at the exit.

Several methods have been developed for introducing the tracer material into the
vessel, but the two most important are

1) continuous addition of tracer to the incoming stream, started at a particular time
   (step input); and
2) addition of a quantity of tracer over a short time interval, the duration of which
   is negligible in comparison with the mean residence time of fluid in the vessel
   (pulse input).

3.2.1 Continuous Addition of Tracer (Step Input)—F Diagrams

Let us consider a case where material flows continuously through a tundish or,
more generally, a reactor vessel. At a given time, $t = 0$, we start adding a tracer to
the inlet stream to the reactor, at a rate that is maintained constant throughout
the experiment. At the same time, we sample the exit stream continuously and
plot the exit concentration of tracer as a function of time.

The results can be plotted in dimensionless, and therefore more general, form
by using the variables

$$F = \frac{c}{c_i} \qquad \text{to express concentration,}$$

$$\theta = \frac{t}{t_r} \qquad \text{to express time,}$$

where

c_i = concentration of tracer in inlet stream,

c = concentration of tracer in exit stream at time t,

t_r = mean residence time of fluid in vessel.

A plot of F against θ is usually called the F diagram (or F curve) for the system.
The value of F represents the concentration in the exit stream as a fraction of the
inlet concentration of tracer. Some typical F diagrams are shown in Figs. 3.2–3.5,
and will be discussed individually. These diagrams have the common characteristic
that, at time zero, the value of F is also zero, while for large values of θ, F approaches
unity. This, of course, reflects the fact that after a certain period of time, all the fluid
originally in the vessel (prior to $t = 0$) is replaced by "new" fluid having the inlet
tracer concentration.

The shape of the F diagram varies considerably between different systems and
can provide useful information as to the flow characteristics in a vessel. The F

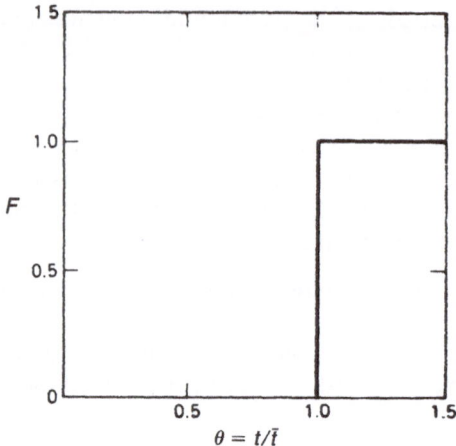

Fig. 3.2. F diagram for step input change for plug flow.

diagrams of some of the most important flow models are discussed in the following paragraphs.

Plug (or Piston) Flow

Figure 3.2 shows the F diagram for a fluid flowing through a vessel in plug flow or piston flow. Under this condition, all fluid elements travel through the vessel at nearly the same speed and preserve their identity (i.e., they do not mix) during their passage.

The step change in the tracer concentration at the inlet at time $\theta = 0$ (i.e., $t = 0$) is reproduced exactly at the exit at time $\theta = 1$ (i.e., $t = t_r$). Thus, the tracer arrives at the exit point at the "expected time," which indicates there is no spread of residence times and all fluid elements spend the same length of time in the vessel.

Although the ideal of plug flow cannot be attained exactly, the flow through long pipes or very long packed beds may be regarded as a sufficiently close approximation.

Backmix (or Perfectly Mixed) Flow

The F diagram of Fig. 3.3 illustrates the behavior of a flow system, which is usually described as backmix flow. In this case, the tracer introduced at the inlet is dispersed instantaneously and uniformly throughout the system; thus, at any time the tracer concentration at the exit must equal that in vessel. It is seen that the tracer appears at the exit immediately after it has been introduced into the inlet stream, but the approach to $F = 1$ is comparatively slow. Thus, the behavior of the tracer indicates that a fraction of the material is retained in the vessel for periods much greater than t_r, whereas other parts of the stream pass through it very rapidly.

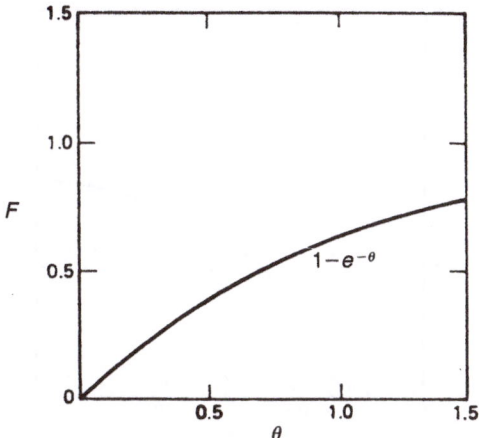

Fig. 3.3. F diagram for backmix flow vessel.

Under conditions of perfect mixing, a material balance on the tracer yields

$$\begin{array}{c}\text{rate of change of}\\ \text{concentration in}\\ \text{vessel}\end{array} = \begin{array}{c}\text{amount of tracer}\\ \text{in inlet stream}\end{array} - \begin{array}{c}\text{amount of tracer}\\ \text{in exit stream}\end{array}$$

$$\bar{V}\frac{dc}{dt} = vc_i - vc, \tag{3.9}$$

where

$\bar{V} \equiv$ volume of fluid in vessel,

$v \equiv$ volumetric rate of flow,

$c_i \equiv$ inlet concentration of tracer,

$c \equiv$ concentration of tracer in vessel and also in the exit stream.

It can be seen that Eq. (3.9) has the form of a first-order rate equation. By integrating this equation, we obtain the following equation for the F diagram of Fig. 3.3:

$$F_t = 1 - e^{-tv/\bar{V}}, \tag{3.10}$$

or in dimensionless form

$$F_\theta = 1 - e^{-\theta}, \tag{3.11}$$

where

$$F_\theta = \frac{c \text{ at time } \theta}{c_i}$$

and

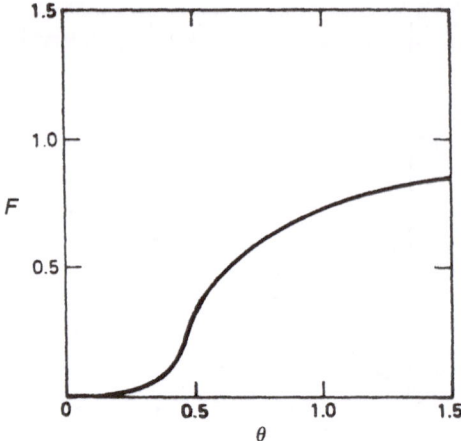

Fig. 3.4. F diagram for combined plug and backmix flow in a vessel.

$$\theta = \frac{t}{t_r}.$$

The idealized concept of perfect mixing may be applied as a good approximation to mechanically agitated vessels. In many metallurgical furnaces, a high degree of mixing may be brought about in the bath by natural convection.

Plug Flow with Axial Mixing

Figure 3.4 represents an intermediate case between ideal models of plug flow and backmix flow. It is seen that the tracer front arrives at the exit point sometime before $t = t_r$, but then there is a rapid increase in the value of F, which reaches unity at $t > t_r$.

Dead Volume Regions

The situation depicted in Fig. 3.5, that is, a rapid rise in the F curve for $t < t_r$, followed by a substantial decrease in slope (still in the $t < t_r$ region) and the subsequent very slow approach to unity at $t > t_r$, indicates there are "dead volume" or inactive zones within the vessel. Thus, the bulk of the fluid spends less time in the vessel than that indicated by the average residence time, whereas a portion is retained for periods much longer than t_r. This type of behavior indicates inefficient use of the reactor volume and should be avoided, or at least minimized, in practice.

In practical situations, dead volume zones may occur owing to eddies trapped at sharp corners, especially when material is introduced or withdrawn through small ports in a large vessel, as illustrated in Fig. 3.6.

In tundishes dead volumes will occur in corners, and also in the lee of baffles or weirs, as shown in Fig. 3.7.

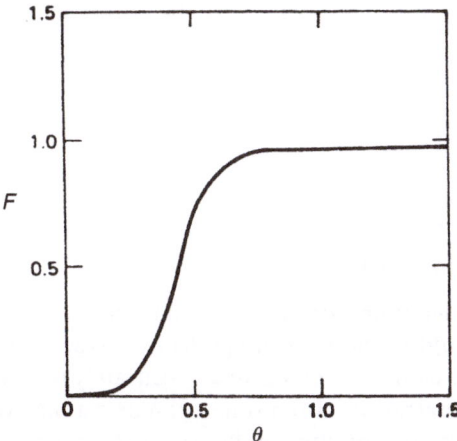

Fig. 3.5. *F* diagram indicating presence of dead volume regions.

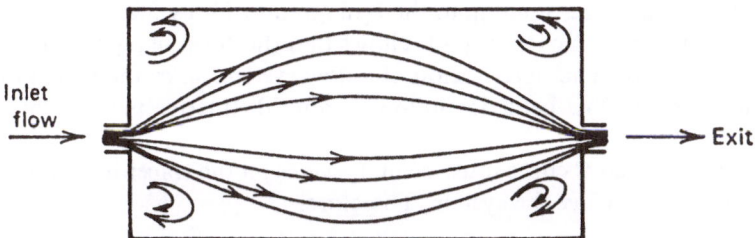

Fig. 3.6. Dead volumes due to eddies trapped at sharp corners.

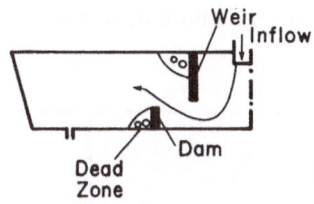

Fig. 3.7. Dead volume formation in lee of flow control devices.

3.2.2 Instantaneous Addition of Tracer (Pulse Input)—*C* Diagrams

The use of step input tracer tests is not practical in many cases for a number of reasons: for example, extensive contamination of product, and cost of tracer. An alternative tracer technique, which is used much more frequently, is the pulse input method. In this case, we introduce a quantity Q of tracer over a time period that is very small in comparison with t_r.

As in the previous case, the concentration of the tracer is determined in the exit stream, and the results are plotted in the form of the dimensionless exit concentration:

$$C = \frac{c}{Q/\bar{V}},\qquad (3.12)$$

where

Q = quantity of tracer injected,

$\bar{V}$ = volume of fluid in vessel.

It can be seen that the denominator of Eq. (3.12) represents the average concentration that the tracer would reach if it were perfectly mixed with the contents of the vessel. In cases where Q cannot be determined accurately (e.g., in radioactive tracer tests), the average concentration of the tracer can be estimated from the weighted average of the exit concentration during the test. For a more detailed treatment of this, the reader is referred to the text by Szekely and Themelis [3].

The results of pulse input tracer tests are plotted in the form of C against dimensionless time ($\theta = t/t_r$). These curves are usually called C diagrams, and some representative plots corresponding to the F diagrams of Figs. 3.2–3.5 are shown in Figs. 3.8–3.11. Inspection of these plots shows that this form of representation may be more useful for the characterization of a flow system, since the C diagrams for the various types of flow behavior follow a more distinctive pattern than the F diagrams.

The area under each C curve must be unity, since all the tracer introduced at the inlet must eventually leave the system. Therefore,

$$\int_0^\infty C\,d\theta = 1.\qquad (3.13)$$

In the idealized case of plug flow, the fluid element containing the tracer material will not mix at all with the rest of the fluid as it travels toward the exit of the vessel (Fig. 3.8).

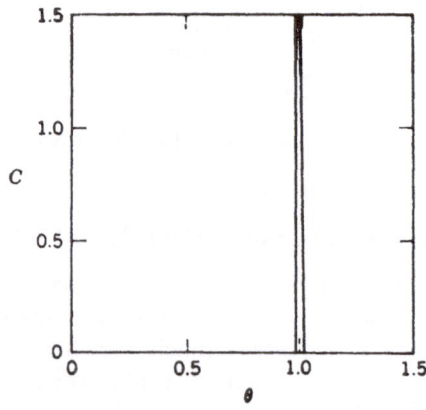

Fig. 3.8. C diagram for pulse input in plug flow vessel.

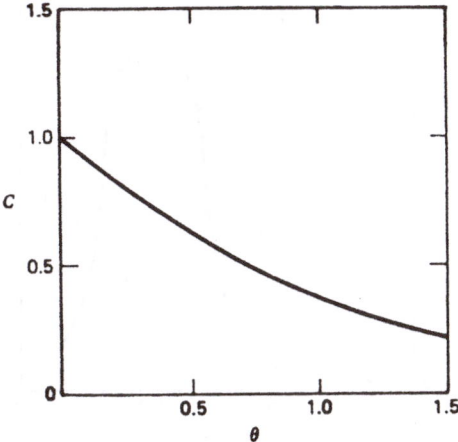

Fig. 3.9. C diagram for backmix flow vessel.

For a perfectly mixed system, the C diagram (Fig. 3.9) exhibits a gradual decrease with time from an initial value of unity. Following the reasoning that led to Eqs. (3.10) and (3.11) for the F diagram of backmix flow, we can easily show that the C curve is described by the following exponential decay equation:

$$C_\tau = \frac{c}{Q/V} = e^{-tv/\bar{V}},$$ (3.14)

or in dimensionless form

$$C_\theta = e^{-\theta}.$$ (3.15)

Comparison of Eqs. (3.11) and (3.15) shows that the C curve is, in fact, a derivative of the F curve. It can be shown that this a general rule that is generally valid regardless of the flow pattern. Therefore, we may write

$$C_\theta = \frac{Q}{V}\frac{\partial F_\theta}{\partial \theta}.$$ (3.16)

In the case of piston flow with a small amount of axial mixing, the C curve exhibits a maximum peak at $\theta = 1$ (Fig. 3.10). It will be shown subsequently that the spread of the curve about the median $\theta = 1$ can be used for quantitative assessment of the mixing conditions in the system.

The presence of dead volume regions (Fig. 3.11) is indicated by a maximum in the C curve, occurring at $t < t_r$; at the peak, we also have $C > 1$. We shall show subsequently that the value and location of this peak may be used to estimate the extent of the dead volume zone in a furnace.

To summarize, we have shown that by the graphical representation of the tracer results we may obtain a good qualitative indication of the behavior of the system; however, for many applications it is desirable to obtain a quantitative measure of

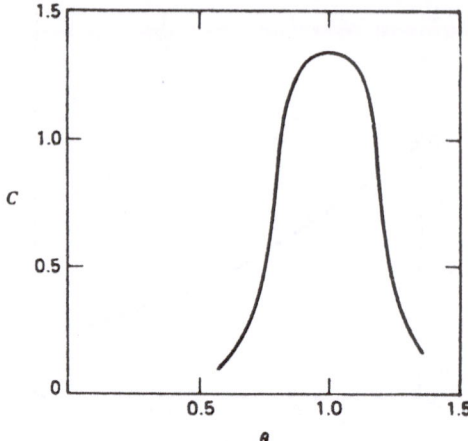

Fig. 3.10. C diagram for plug flow with axial mixing.

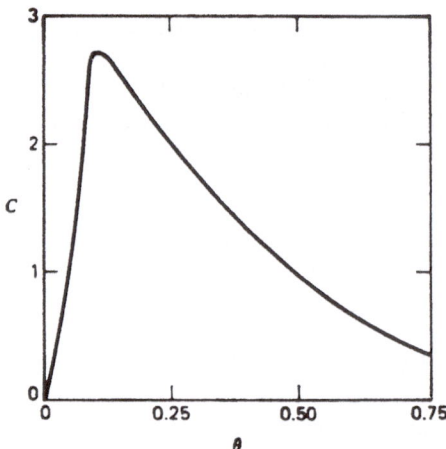

Fig. 3.11. C diagram of plug and backmix flow in presence of dead volume regions.

mixing or departure from ideality. The techniques available for making quantitative deductions from the C and F diagrams are discussed in the next section.

While a rigorous representation of tracer dispersion is possible through the solution of the convective mass transfer equation, which will be discussed subsequently, much can be learned by considering an approximate representation of the system as follows: We shall assume that the system will consist of three regions, namely, a plug flow region, a perfectly mixed region, and a dead volume region, such as sketched in Fig. 3.12.

The usefulness of the mixed model lies in the fact that the experimentally determined C diagram for a particular system can be analyzed to determine the relative

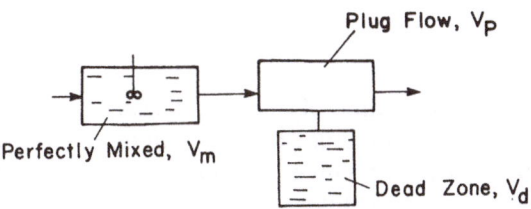

Fig. 3.12. Three selected regions of flow in a tundish system.

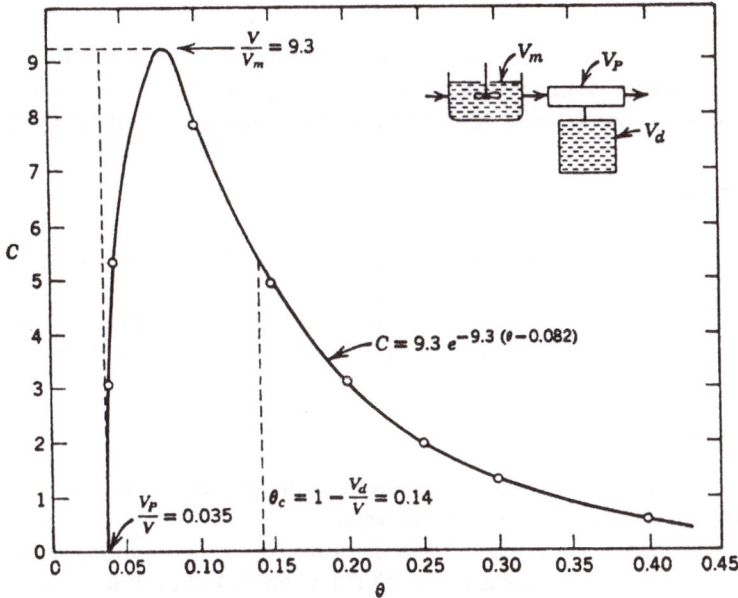

Fig. 3.13. Analysis of C curve according to the mixed model [2].

volumes of plug flow, backmix flow, and dead volume regions. In this manner, the flow conditions in the system can be defined quantitatively and compared with C diagrams obtained after instituting the necessary operational changes (e.g., changes in flow geometry or introduction of dams or agitators). As an illustration of the use of this model, let us consider the experimentally determined C diagram of Fig. 3.13. The tracer material first appears at the exit of the vessel at time t_p. By dividing t_p by the calculated average residence time of fluid in the vessel, we obtain the fraction of reactor volume that can be assumed to be in plug flow:

$$\theta_p = \frac{t_p}{t_r} = \frac{V_p}{V}, \tag{3.17}$$

where V_p is the plug flow volume and $\bar{V}$ is the total volume of fluid in the tundish.

Figure 3.13 shows that the maximum tracer concentration is observed at time $t/t_r < 1$ and has a value

$$C_{\text{max}} = \frac{c_{\text{max}} \, \bar{V}}{Q} > 1, \qquad (3.18)$$

where c_{max} is the peak concentration of tracer at the exit.

This behavior indicates the presence of dead volume regions in the vessel. To determine the fraction of dead volume, we must first calculate the actual mean residence time of liquid in the reactor. We then define the dimensionless mean residence time as follows:

$$\theta_{\text{mean}} = \frac{\text{Actual mean residence time}}{\text{Calculated mean residence time}}. \qquad (3.19)$$

It is obvious that the actual mean residence time will differ from the calculated mean [i.e., (reactor volume)/(volumetric flow rate)] only in the presence of dead volume regions. It can be readily shown that the following relationship applies:

$$\frac{V_d}{V} = 1 - \theta_{\text{mean}} \qquad (3.20)$$

where V_d is the volume of the dead regions in the vessel. Equation (3.20) shows that in the absence of dead volume ($V_d = 0$), $t_{\text{mean}} = t_r$.

Once we have obtained the value of the plug flow volume [Eq. (3.17)] and the dead volume region [Eq. (3.20)], we can calculate the volume of the backmix flow from the equation

$$V_p + V_d + V_m = \bar{V}, \qquad (3.21)$$

that is, the sum of the plug flow, dead volume, and perfect mixing regions must be equal to the total volume of fluid in the vessel.

The backmix flow volume may also be determined from the value of the maximum concentration, which, as shown in Eq. (3.18), is greater than unity. The lesser the degree of mixing in the vessel, the higher will be the value of the peak concentration. The following relationship applies for a mixed model:

$$C_{\text{max}} = \frac{\bar{V}}{v_m}. \qquad (3.22)$$

A "reliability" test for a mixed model is to compute the value of V_m from Eq. (3.22) and see whether it matches the value obtained by the difference from the computed V_p and V_d values [Eq. (3.21)].

As a matter of fact, most tundishes feeding a single. (i.e., double-, if symmetry is allowed for) strand casting system may be described in terms of such a mixed model, at least in an appropriate fashion. In general, there is a relatively small plug flow region, that is, the breakthrough time is quite short, and a significant fraction of the total volume is dead or inactive. Finally, there is a significant volume that is well mixed; this usually corresponds to the pouring box. This approximate method of analysis is helpful in providing a preliminary assessment of tundish performance.

Let us conclude this section with the presentation of a selection of tracer results obtained both in water models and in real tundish systems.

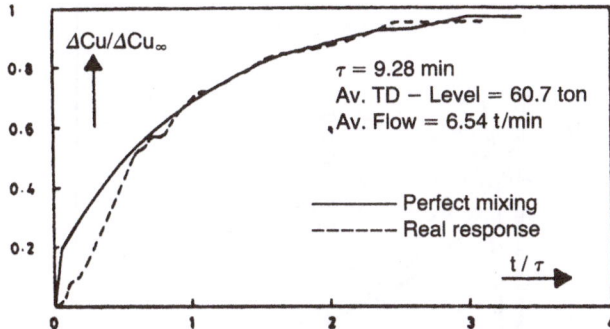

Fig. 3.14. Step response of a tundish without obstacles; Cu (tracer): 0.012–0.07% [4].

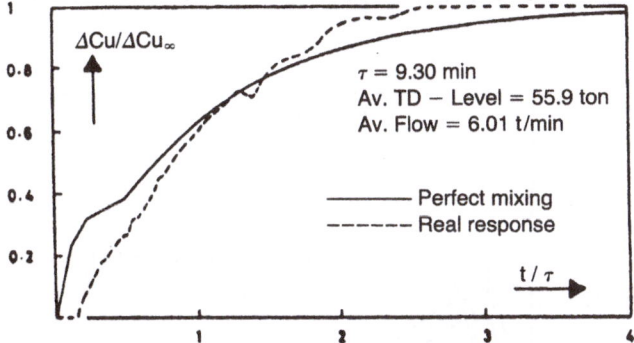

Fig. 3.15. Response of a tundish with weirs and dams; Cu (tracer): 0.01–0.07% [4].

As a practical matter, when using water models, there is a great variety of possibilities for tracer use, including dyes, acids, or bases. When using tracers in steel mill practice, one could use radioactive materials (less popular) or some other readily soluble metal, such as copper, which can be easily analyzed.

Figures 3.14 and 3.15, taken from the work of van der Heiden et al. [4], show typical F curves obtained for a real tundish. These indicate in a qualitative sense that the use of dams or weirs will improve matters, since they will provide for a finite breakthrough time. It should be noted that both these plots would seem to indicate that the systems behave almost like a perfectly mixed vessel. This is somewhat fortuitous, because in fact, the system is not completely mixed, but is some combination of a small plug flow region and a significant dead volume, in addition to having a well-mixed portion.

Examination of Figs. 3.14 and 3.15 also shows that the F curves are not ideal for descriminating between different types of tundish behavior.

Figures 3.16–3.19 show sets of C curves, that is, corresponding to stepwise addition of the tracer, as reported by Martinez et al. [5]. These curves are highly instructive, because they clearly show an excellent correspondence between the real

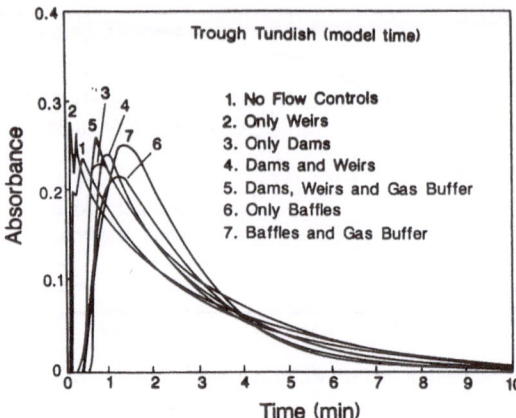

Fig. 3.16. Trough tundish residence time diagrams (one-quarter scale model) [5].

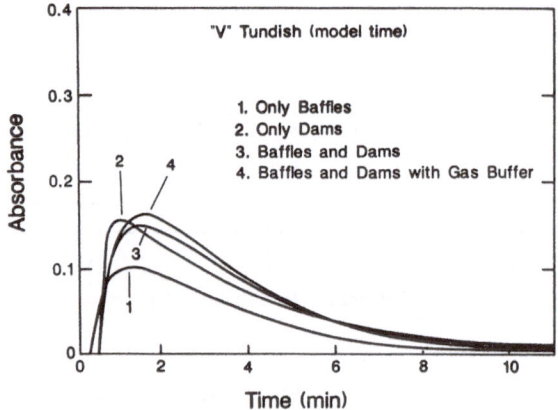

Fig. 3.17. Residence time diagrams for the "V" tundish (one-quarter scale model) [5].

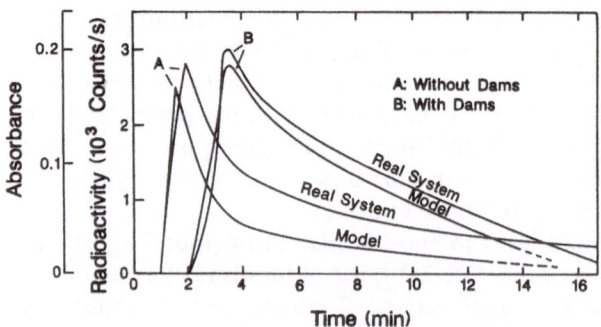

Fig. 3.18. Residence time diagrams for real system and model (one-third scale) [5].

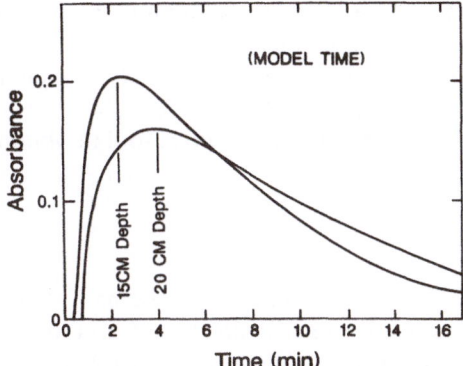

Fig. 3.19. Residence time diagrams for the one-third scale model with different tundish depths [5].

system and its water model. These plots also show a very marked departure from a "perfectly mixed" system in all cases.

The quite dramatic effect of using different tundish configurations and inserts is readily apparent on examining these plots, clearly illustrating the important use of tundish modeling and tracer studies as effective tools in optimizing tundish performance.

One should stress, however, that water models of tundishes in conjuction with tracers are excellent diagnostic tools for identifying good and poor tundish designs; however, in the absence of detailed velocity measurements, these techniques fall somewhat short of providing the necessary insight into the detailed working of tundishes.

Furthermore, the physical modeling approaches are of limited use in addressing the questions of inclusion flotation, heat transfer, and mass transfer, because these phenomena cannot be readily modeled.

3.3 Summation

In this chapter, we have introduced the reader to the concepts of physical modeling and tracer techniques.

Physical Modeling

We have shown that due to the form of the Navier–Stokes equations, one can represent or model the many aspects of tundish behavior by using water as the model fluid, provided key dimensionless numbers such as the Froude number and Reynolds number are identical, in addition to having the model and the prototype of the same shape (geometric similarity).

For practical tundish systems, this means that the model and the prototype have to be of approximately the same dimension.

As an important caveat, once we start examining heat transfer and multiphase situations such as the behavior of inclusions or entrainment of gas streams, a number of additional dimensionless numbers will enter the picture, and the modeling exercise will become much less exact.

Nevertheless, water modeling is a very effective tool in obtaining a better understanding of tundish behavior.

Tracer Studies

Once a physical model has been constructed, the easiest diagnostic tool to be employed is the use of tracers. We have defined the F and the C diagrams, and have shown how these may be interpreted. We have shown that the use of tracers is very effective in identifying by-pass streams and dead zones; at the same time, while the tracers are very helpful in identifying some gross features of tundish systems, they will not provide the detailed insight that might be needed for improved tundish design and operation. More specifically, the results of a tracer test will point to the symptom of poor performance, but cannot provide a guide as to how this performance may be improved. For this, we needed mathematical modeling tools that were described in Chapter 2.

References

1. A. Mclean, Course Notes on Tundish Metallurgy for Iron and Steel Society, Toronto, 1987.
2. N.J. Themelis and P. Spira, Trans. AIME, *236*, 821 (1964).
3. J. Szekely and N.J. Themelis, Rate Phenomena in Process Metallurgy, Wiley-Interscience, New York (1971).
4. A. van der Heiden, P.W. van Hasselt, W.A. de Jong, and F. Blaas, "Inclusion Control for Continuously Cast Products," Proceedings of the 5th International Iron and Steel Congress, Washington, DC, pp. 755–760 (1986).
5. E. Martinez, M. Maeda, L.J. Heaslip, G. Rodriguez, and A. McLean, "Effects of Fluid Flow on the Inclusion Separation in Continuous Casting Tundish," Trans. I.S.I.J., *26*, 724 (1986).

4 Computed Results on Tundish Systems

In the previous chapters we discussed the principles of mathematical and physical modeling as applied to tundish operations and have also shown some selected physical modeling results. In this chapter we shall develop a general mathematical formulation describing the flow behavior of tundish systems and then will present a selection of computed results. These results were so chosen as to provide a good overall insight into tundish behavior. Regarding the organization of this chapter, Section 4.1 will contain the general formulation; the computed fluid flow results, including tracer dispersion and inclusion behavior, will be presented in Section 4.2; the behavior of nonisothermal tundishes will be described in Section 4.3; and the effect of externally imposed magnetic fields will be given in Section 4.4. Shallow tundishes will be described in Section 4.5, while wave formation in tundish systems will be treated in Section 4.6.

4.1 General Formulation of Flow Phenomena in Tundishes

4.1.1 Fluid Flow

Figure 4.1 is a schematic sketch of a tundish system. As we discussed, the tundish may be regarded as a trough into which molten metal is being poured and from which molten metal is being discharged. The flow will be turbulent in the inlet and outlet regions and, in all likelihood, transitional in the bulk of the tundish system.

In order to render the problem mathematically manageable, at least in the first instance, let us make the following simplifying assumptions:

1) The flow is macroscopically steady; that is, we shall ignore the effect of a ladle change.
2) We shall assume that the free surface is flat.
3) We shall neglect any air or gas entrainment by the incoming metal stream.

It is noted that the first two of these assumptions will be relaxed in Section 4.6.

Within this framework, the flow behavior of the system (including both conventional and shallow tundishes) may be represented by the continuity and the three components of the Navier–Stokes equations, which take the following forms:

Continuity:

$$\frac{\partial U_x}{\partial x} + \frac{\partial U_y}{\partial y} + \frac{\partial U_z}{\partial z} = 0; \tag{4.1}$$

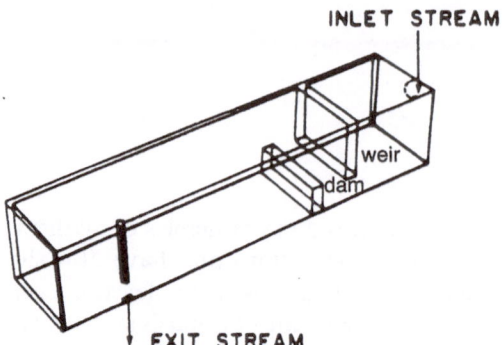

Fig. 4.1 Typical single-strand tundish arrangement.

Momentum in x-direction:

$$\rho\left(U_x\frac{\partial U_x}{\partial x} + U_y\frac{\partial U_x}{\partial y} + U_z\frac{\partial U_x}{\partial z}\right) = \mu_{\text{eff}}\left(\frac{\partial^2 U_x}{\partial x^2} + \frac{\partial^2 U_x}{\partial y^2} + \frac{\partial^2 U_x}{\partial z^2}\right) - \frac{\partial P}{\partial x}; \tag{4.2}$$

Momentum in y direction:

$$\rho\left(U_x\frac{\partial U_y}{\partial x} + U_y\frac{\partial U_y}{\partial y} + U_z\frac{\partial U_y}{\partial z}\right) = \mu_{\text{eff}}\left(\frac{\partial^2 U_y}{\partial x^2} + \frac{\partial^2 U_y}{\partial y^2} + \frac{\partial^2 U_y}{\partial z^2}\right) - \frac{\partial P}{\partial y}; \tag{4.3}$$

Momentum in z direction:

$$\rho\left(U_x\frac{\partial U_z}{\partial x} + U_y\frac{\partial U_z}{\partial y} + U_z\frac{\partial U_z}{\partial z}\right) = \mu_{\text{eff}}\left(\frac{\partial^2 U_z}{\partial x^2} + \frac{\partial^2 U_z}{\partial y^2} + \frac{\partial^2 U_z}{\partial z^2}\right) - \frac{\partial P}{\partial z}; \tag{4.4}$$

μ_{eff} is the "effective viscosity" defined in Eq. (2.5).

It should be stressed that the nature of tundish flows is inherently three dimensional, so that the consideration of two-dimensional slices would serve only as a highly qualitative indication.

The effective viscosity appearing in Eqs. (4.2)–(4.4) has to be represented in terms of a turbulence model. In the present case, we shall use the version of the K-ε model of Launder and Spalding [1] in which K is the specific turbulence energy [Eq. (2.14)] and ε is the rate of turbulence energy dissipation. The associated relationships and the method of calculating K and ε have been detailed in Section 2.1.

The boundary conditions will have to express the following:

1) Velocity of the inlet stream has to be specified.
2) A no-slip condition has to be prescribed at the solid surfaces. However, in order to preclude the need to calculate right to the wall where flow properties are steep, the velocities adjacent to the wall are located in the turbulent region outside the viscous sublayer. A logarithmic wall function is then applied to bridge the intervening layer such that the wall shear stress τ_w, is calculated from the relationship

$$\frac{U}{(\tau_w/\rho)^{1/2}} = \frac{1}{\kappa} \ln\left(\frac{E y_p (\tau_w \rho)^{1/2}}{\mu_l}\right), \tag{4.5}$$

in which κ is the von Karman constant, E is a sublayer resistance factor, and y_p is the perpendicular distance of the near-wall grid node to the wall.

The boundary condition on K is obtained by assuming that this stress is constant across the viscous sublayer from which we obtain the relationship:

$$K = \frac{\tau_w}{\rho c_\mu^{1/2}}, \tag{4.6}$$

where c_μ is an empirical constant.

Finally, by assuming that the rate of dissipation of turbulence energy is equal to the rate of production in this region (convection and diffusion of energy being small), we obtain the following relationship that serves as the boundary condition on ε; thus,

$$\varepsilon = \frac{\tau_w^{1.5}}{\rho^{1.5} \kappa y_p}. \tag{4.7}$$

Details of the preceding approach and the values of the various constants are contained in Launder and Spalding [2].

3) Zero shear stress is specified at the free surface. This assumption may not always be valid, because if a viscous slag layer is provided to cover the melt, then a zero velocity at the top surface would be a better assumption. However, since the velocities tend to be quite small in the bulk, this assumption is unlikely to make a significant difference.

4) We note that while it might be tempting to prescribe the melt velocity at the exit nozzles, this would in fact, overspecify the problem. We need to specify the pressure at the outlet and then check the internal consistency of the results by ensuring that an overall mass balance is indeed being satisfied.

4.1.2 Tracer Dispersion

Tracer dispersion is an important tool in tundish characterization. As discussed in Chapter 3, tracer dispersion may be represented by

$$\rho\left(\frac{\partial C}{\partial t} + U_x \frac{\partial C}{\partial x} + U_y \frac{\partial C}{\partial y} + U_z \frac{\partial C}{\partial z}\right) = D_{\text{eff}}\left(\frac{\partial^2 C}{\partial x^2} + \frac{\partial^2 C}{\partial y^2} + \frac{\partial^2 C}{\partial z^2}\right), \tag{4.8}$$

in which D_{eff} is an "effective diffusivity" defined as

$$D_{\text{eff}} = D + D_t, \tag{4.9}$$

where D is the laminar diffusivity and D_t is the eddy diffusivity. The latter is related to the turbulent viscosity through the turbulent Schmidt number as

$$\frac{\mu_t}{\rho D_t} = 1. \tag{4.10}$$

A zero-flux condition is imposed on Eq. (4.8) on all solid surfaces, symmetry planes, and free surface.

4.1.3 Inclusion Behavior

The precise description of the behavior of inclusions in tundish systems would be a very complex task indeed, because inclusion particles may

coalesce in the highly turbulent entry region;
become attached to the walls, dams, and weirs; or
float out of the system.

In the present case, we shall address only the flotation of spherical inclusion particles. Nevertheless, even this simplified treatment should provide helpful guidance regarding the overall behavior of these systems.

Following an approach similar to that of Tacke and Ludwig [3], we assume that the transport of inclusions would be governed by an equation system similar to Equ. (4.8) but without the transient term, since the motion of the particles will be inherently steady in a time-averaged statistical sense. The concentration C is thus defined as the number of particles per unit volume. We also neglect the contribution of the diffusion term in the equation.

The effect of buoyancy is modeled by assuming that the particle has constant terminal rising velocity U_T which, for a spherical particle in a stagnant fluid is given by the Stokes law as

$$U_T = \frac{2R_p^2(\rho_p - \rho_f)g}{9\mu},\tag{4.11}$$

where

$R_p \equiv$ particle radius,

$\rho_p \equiv$ particle density,

$\rho_f \equiv$ fluid density,

$\mu \equiv$ laminar viscosity.

This velocity is added vectorially to the vertical component of velocity U_z in the transport equation.

The boundary conditions specify zero flux at all solid surfaces. Ideal absorption conditions are assumed at the free surface such that the flux q_s of particles there is governed by the relationship

$$q_s = U_T C_s,\tag{4.12}$$

where C_s is the particle concentration at the surface.

4.1.4 Heat Transfer

The equation of conservation of energy takes the form

$$\rho\left(\frac{\partial T}{\partial t} + U_x\frac{\partial T}{\partial x} + U_y\frac{\partial T}{\partial y} + U_z\frac{\partial T}{\partial z}\right) = \frac{\mu_{\text{eff}}}{\sigma_T}\left(\frac{\partial^2 T}{\partial x^2} + \frac{\partial^2 T}{\partial y^2} + \frac{\partial^2 T}{\partial z^2}\right) + q_v,\tag{4.13}$$

where σ_T is the turbulent Prandtl number (≈ 1) and q_v is the volumetric heat generation, which is zero in the absence of auxiliary heating.

All walls are assumed to be thermally insulated while the top surface is allowed to loose heat by conduction through the slag and subsequently by Stefan–Boltzmann radiation to the surroundings. Details of this treatment can be found in Chapter 10 of Szekely and Themelis [4].

4.1.5 Solution Procedure

Equations (4.1)–(4.13) represent the complete description of the problem. The turbulent Navier–Stokes equations are nonlinear, so that numerical methods have to be used in order to obtain a solution.

At the present time, there are several commercially available software packages capable of solving this problem, and these have been listed in Table 2.1. While these codes are readily available, at the present time one would still require a skilled numerical analyst and a person with a good knowledge of fluid mechanics in order to generate meaningful results with confidence. Furthermore, there is a critical need for the independent evaluation of the computed results, through order-of-magnitude checks and other forms of consistency analyses.

The principal practical problems that may arise include the following:

lack of convergence;
difficulties in representing complex shapes (dams, weirs, inclined walls, angled slots, etc.); and
disparities in the various scales, viz., diffusion, viscous transport, etc.

The results that will be presented in the following sections were obtained using PHOENICS code [5]. This code uses a finite-domain, fully implicit, iterative procedure that derives from the SIMPLE algorithm of Patankar and Spalding [6].

Before any set of results was regarded as acceptable, several consistency checks were made, including the verification that the results were indeed grid independent and that the overall conservation of mass and of the tracer was indeed satisfied to within 0.1% of the input value. Tracer conservation, because it involved a time scale and due to the "long tail" of the tracer curves, was the most difficult criterion to meet.

Small grid spacings were used near solid surfaces and other regions where flow properties have steep gradients. Typically, a $25 \times 11 \times 15$ nonuniform grid structure in the longitudinal, transverse, and vertical directions, respectively, were found to be grid independent for one-quarter of the tundish. The calculation typically took about 12 h on a Microvax II digital computer.

In the remainder of this chapter, we shall present a selection of the computed results and, in a limited number of cases, a comparison with the experimental measurements.

This material will be divided into the following sections:

4.2 Conventional tundishes—fluid flow.
4.3 Conventional tundishes—heat transfer.
4.4 Conventional tundishes—magnetic flow control.

4.5 Shallow tundish behavior.
4.6 Conventional tundishes—free surface phenomena.

4.2 Conventional Tundishes: Fluid Flow, Tracer Dispersion, and Inclusion Behavior

Figure 4.2, taken from ref. [7], shows a schematic sketch of the "standard" tundish that will be considered in these calculations. It is seen that this configuration consists of the entry nozzle and two exit nozzles. Furthermore, there is provision for up to two weirs and a dam as flow-control devices. The location of flow-control devices is given in Table 4.1 and the principal input parameters are summarized in Table 4.2.

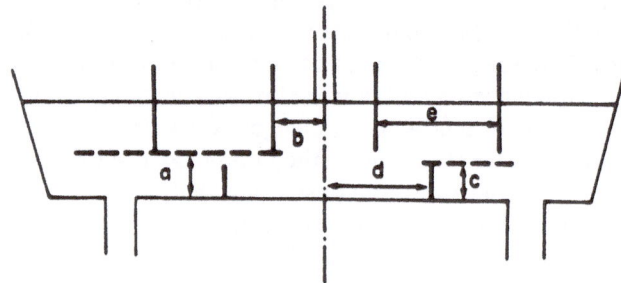

Fig. 4.2. A schematic of a two-strand tundish system after PCH Physico Chemical Hydrodynamics, 9, J. Szekely et al., "The Mathematical Modelling of Complex Fluid Flow Phenomena in Tundishes," 1987, Pergamon Journals Ltd.

Table 4.1. Location of flow-control devices (see Fig. 4.1)

Configuration	a (mm)	b (mm)	c (mm)	d (mm)	e (mm)
1 No flow control	—	—	—	—	—
2 Flow control with one weir, one dam	229	762	229	1067	—
3 Flow control with one weir, one slotted dam	229	762	229	1067	—
4 Flow control with two weirs, one dam	229	762	229	1067	888
5 Flow control with one weir, one dam, and allowance for wall inclination	229	762	229	1067	—

Table 4.2. Principal input parameters of conventional tundishes

Tundish length	6.79 m
Tundish width	0.65 m
Melt depth	0.75 m
Inlet stream velocity	7.67 m/s
Inlet nozzle diameter	54 mm
Outlet nozzle diameter	54 mm
Inclination of walls	8°
Particle density/fluid density	0.5

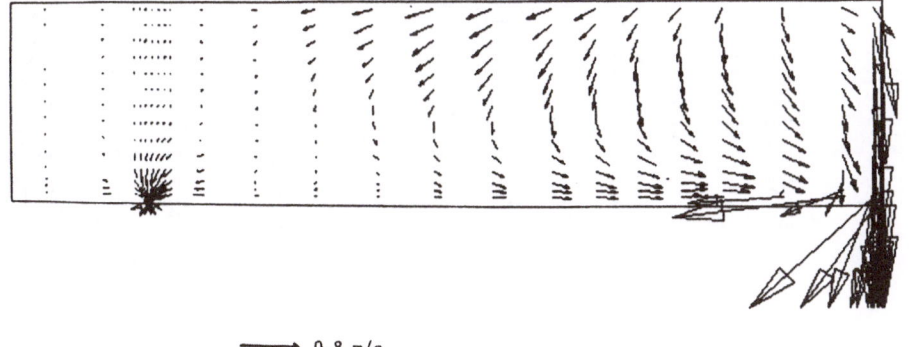

$\longrightarrow$ 0.8 m/s

Fig. 4.3a. Velocity field near the central plane (0.15y) in the absence of flow control.

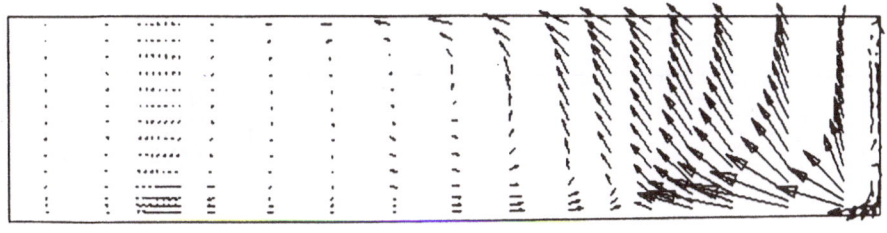

$\longrightarrow$ 0.5 m/s

Fig. 4.3b. Velocity field near the side wall (0.8y) in the absence of flow control.

4.2.1 Computed Results

In the following we shall present a selection of the computed results, with emphasis on data that provide a useful insight into the behavior of the system.

Figures 4.3.a and 4.3b show the velocity maps in the absence of flow control in two vertical slices, one (at 0.15y) near the plane of the inlet stream, and the other (at 0.8y) near the vertical walls. These plots show both the high velocities in the inlet plane, and the marked recirculation involving upward flow near the side walls.

Figures 4.4a and 4.4b show the corresponding plots for flow control, where it is seen that the presence of a dam and a weir will provide for a more regular flow outside the "pouring box." Inspection of these figures also shows the rather stagnant region behind the dam.

Figures 4.5a and 4.5b show the velocity plots for a slotted dam. As expected, there is a significant cross flow, and the tendency for short circuiting near the bottom is readily apparent.

Figures 4.6a and 4.6b show plots for two weirs and one dam, and it is seen that under these conditions the tendency for short circuiting is even more aggravated. Here, flow control may actually be harmful!

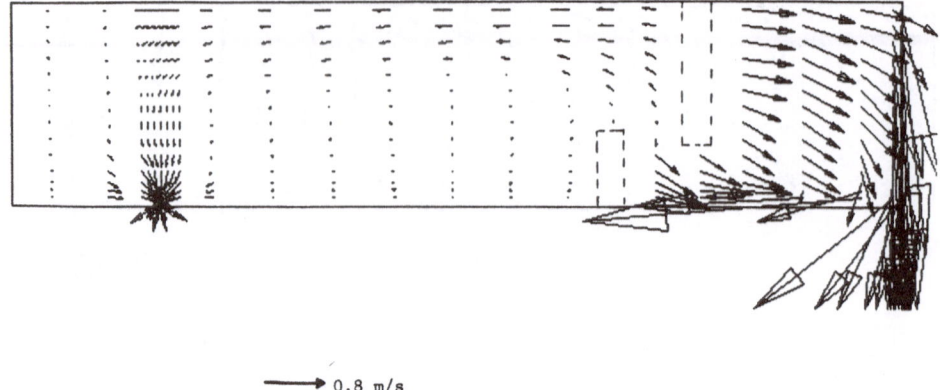

⟶ 0.8 m/s

Fig. 4.4a. Velocity field at 0.15y in the presence of flow control with one dam and one weir.

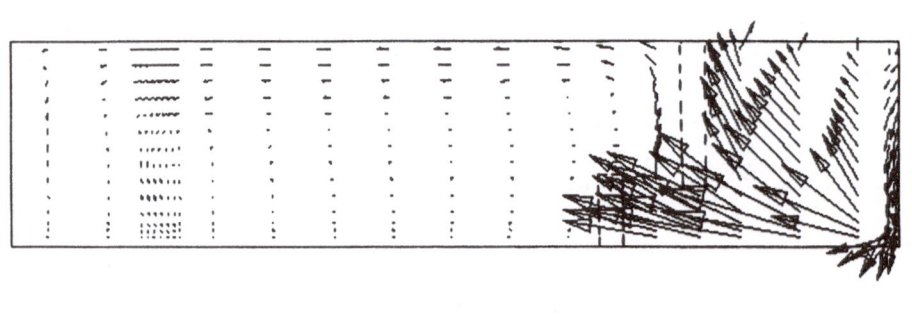

⟶ 0.6 m/s

Fig. 4.4b. Velocity field at 0.8y in the presence of flow control with one dam and one weir.

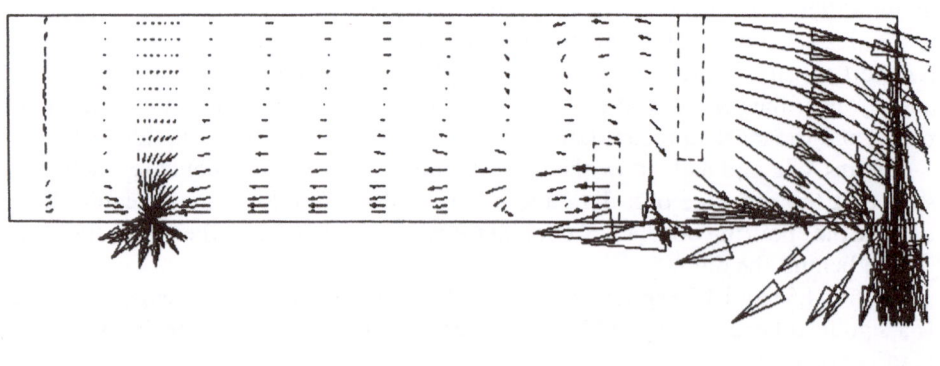

⟶ 0.8 m/s

Fig. 4.5a. Velocity field at 0.15y in the presence of flow control with one slotted dam and one weir.

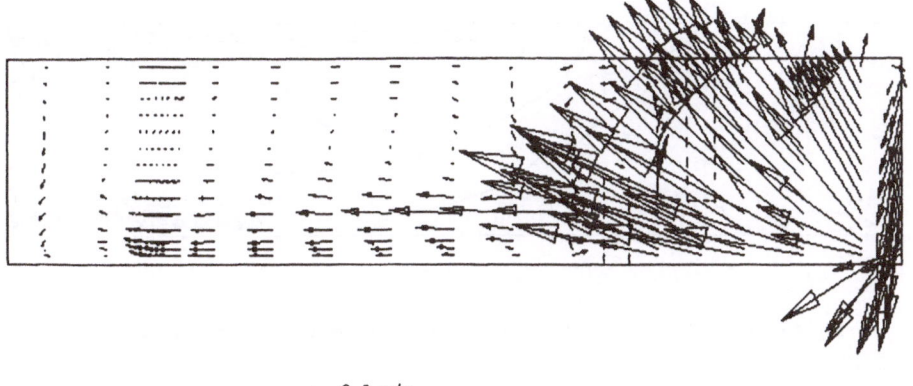

→ 0.3 m/s

Fig. 4.5b. Velocity field at 0.8y in the presence of flow control with one slotted dam and one weir.

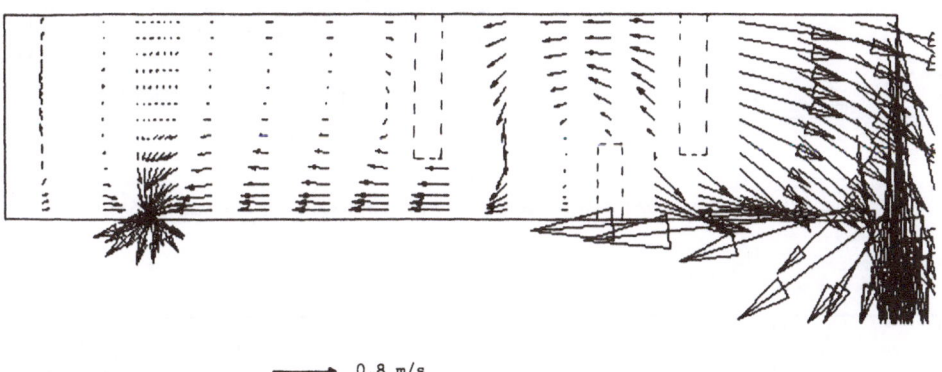

→ 0.8 m/s

Fig. 4.6a. Velocity field at 0.15y in the presence of flow control with one dam and two weirs.

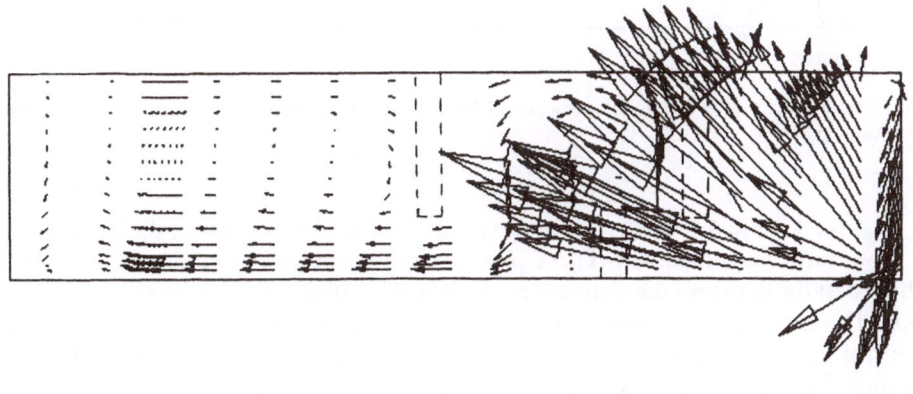

→ 0.3 m/s

Fig. 4.6b. Velocity field at 0.8y in the presence of flow control with one dam and two weirs.

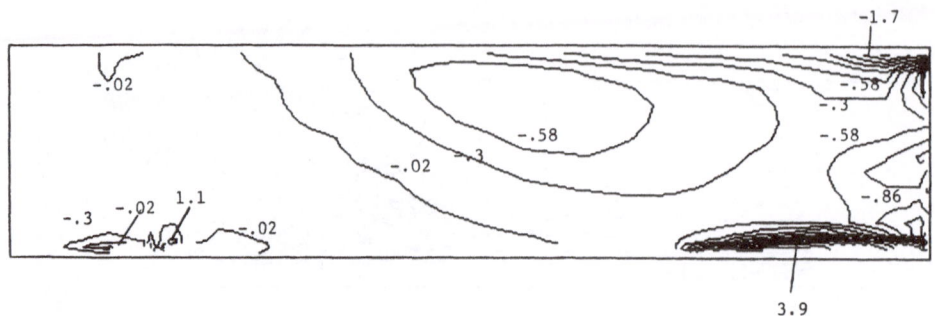

Fig. 4.7a. Vorticity at 0.15y in the absence of flow control. Contour interval = 0.28.

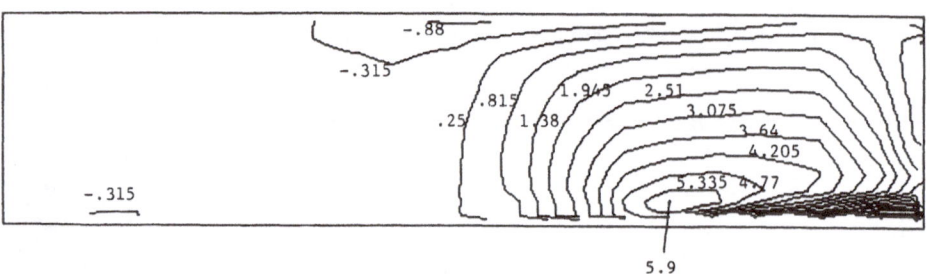

Fig. 4.7b. Vorticity at 0.8y in the absence of flow control. Contour interval = 0.565.

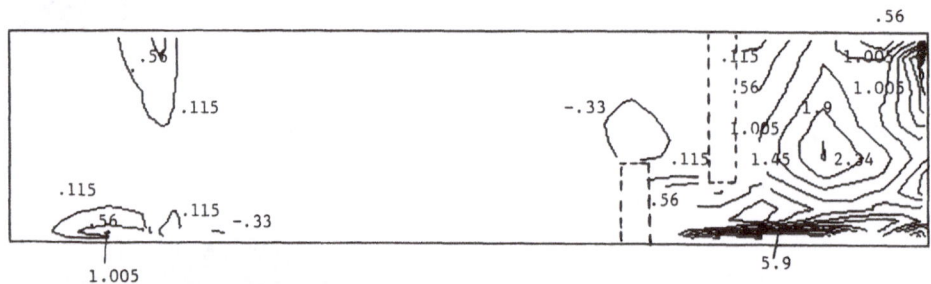

Fig. 4.8a. Vorticity at 0.15y in the presence of flow control with one dam and one weir. Contour interval = 0.445.

Figures 4.7 and 4.8 show the vorticity plots in the absence and in the presence of a flow-control arrangement. These plots are quite instructive, because they clearly show that the dam and the weir arrangement will confine the vorticity to the pouring box. The higher values of the vorticity near the exit nozzle in the absence of flow control clearly indicate that, properly applied, flow control should be helpful in minimizing vorticity and, hence, the tendency for vortex formation.

Figure 4.9 shows the vorticity plots for a slotted dam, and it is seen that the high

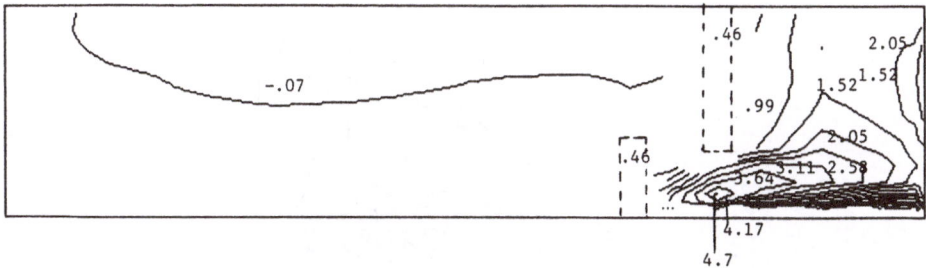

Fig. 4.8b. Vorticity at 0.8y in the presence of flow control with one dam and one weir. Contour interval = 0.53.

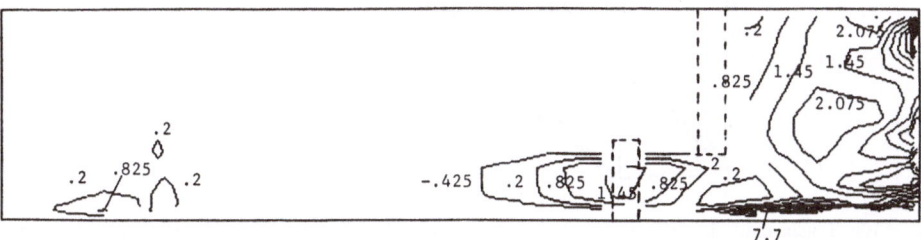

Fig. 4.9a. Vorticity at 0.15y in the presence of flow control with one slotted dam and one weir. Contour interval = 0.625.

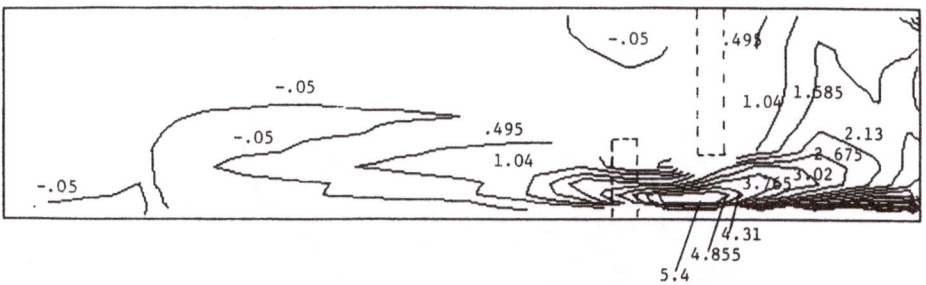

Fig. 4.9b. Vorticity at 0.8y in the presence of flow control with one slotted dam and one weir. Contour interval = 0.545.

velocities near the bottom will give rise to high vorticities, and, hence, to a high tendency for vortexing.

Each of the following figures (4.10–4.12), all taken from Ref. [7], shows the dispersion of a tracer in the form of concentration isopleths at two distinct times. Inspection of these plots shows that in the absence of flow control (Fig. 4.10) the tracer is rapidly dispersed and will start appearing in the outlet stream within about 40 s. The use of a weir and dam arrangement, depicted in Fig. 4.11, shows a marked

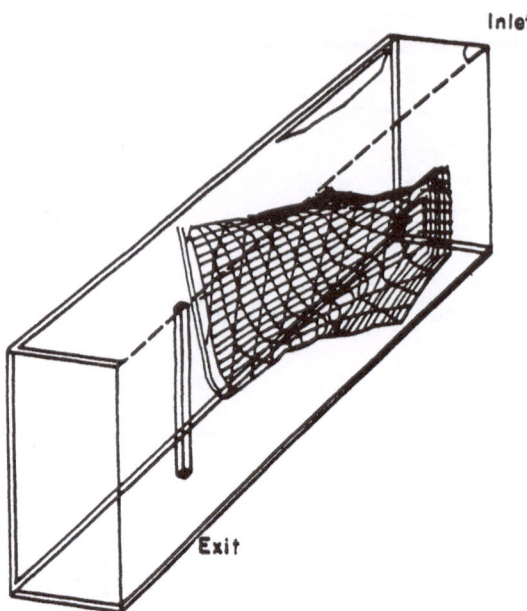

Fig. 4.10a. Tracer map after 10 s for no flow control after PCH Physico Chemical Hydrodynamics, 9, J. Szekely et al., "The Mathematical Modelling of Complex Fluid Flow Phenomena in Tundishes," 1987, Pergamon Journals Ltd.

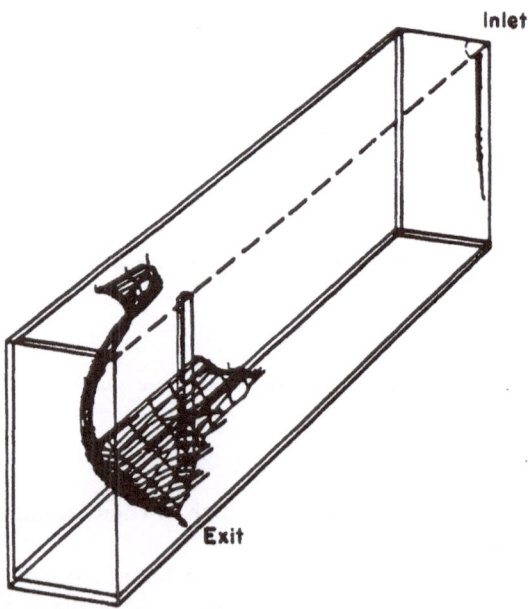

Fig. 4.10b. Tracer map after 40 s for no flow control after PCH Physico Chemical Hydrodynamics, 9, J. Szekely et al., "The Mathematical Modelling of Complex Fluid Flow Phenomena in Tundishes," 1987, Pergamon Journals Ltd.

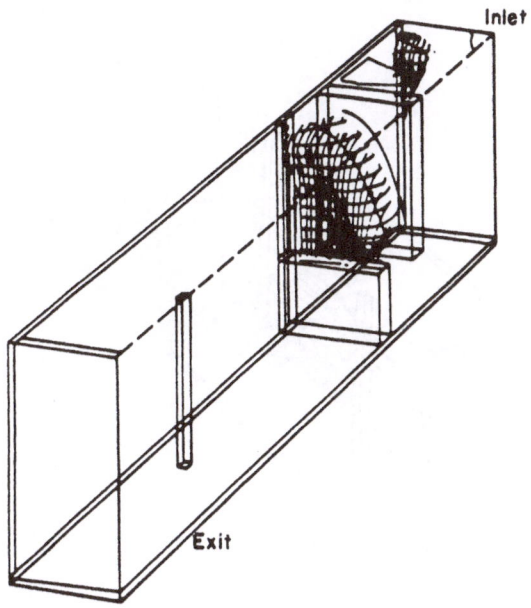

Fig. 4.11a. Tracer map after 10 s for flow control with one dam and one weir after PCH Physico Chemical Hydrodynamics, 9, J. Szekely et al., "The Mathematical Modelling of Complex Fluid Flow Phenomena in Tundishes," 1987, Pergamon Journals Ltd.

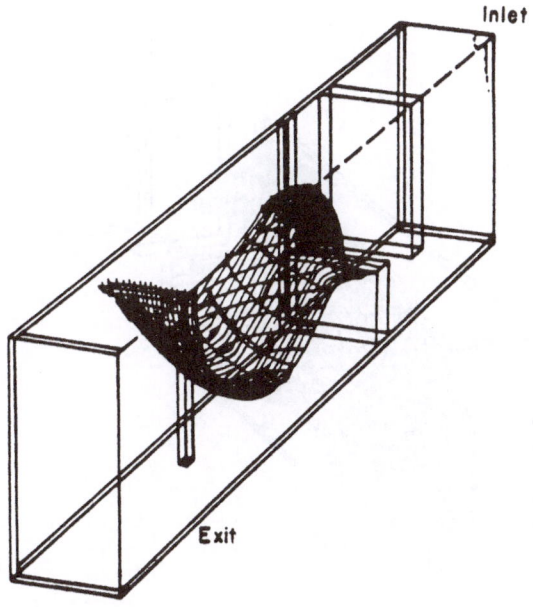

Fig. 4.11b. Tracer map after 40 s for flow control with one dam and one weir after PCH Physico Chemical Hydrodynamics, 9, J. Szekely et al., "The Mathematical Modelling of Complex Fluid Flow Phenomena in Tundishes," 1987, Pergamon Journals Ltd.

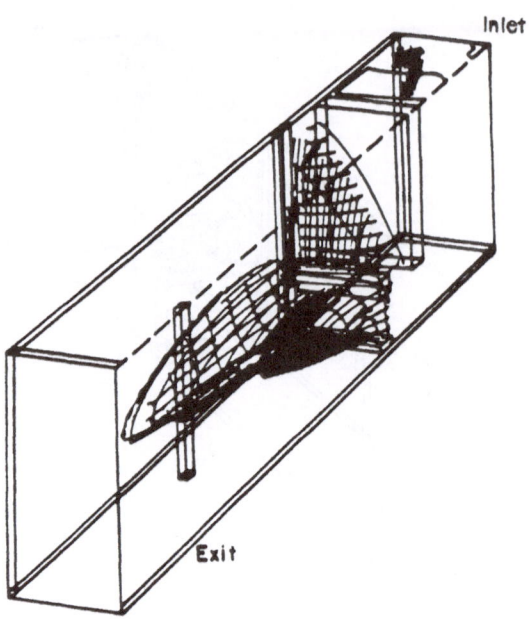

Fig. 4.12a. Tracer map after 10 s for flow control with one slotted dam and one weir after PCH Physico Chemical Hydrodynamics, 9, J. Szekely et al., "The Mathematical Modelling of Complex Fluid Flow Phenomena in Tundishes," 1987, Pergamon Journals Ltd.

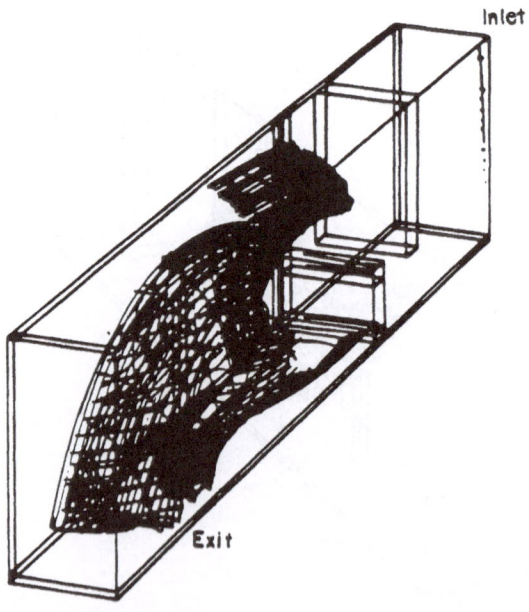

Fig. 4.12b. Tracer map after 40 s for flow control with one slotted dam and one weir after PCH Physico Chemical Hydrodynamics, 9, J. Szekely et al., "The Mathematical Modelling of Complex Fluid Flow Phenomena in Tundishes," 1987, Pergamon Journals Ltd.

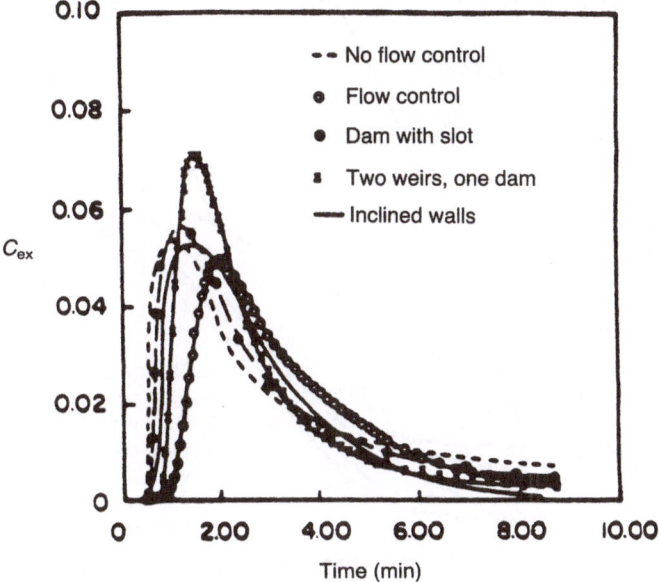

Fig. 4.13. Exit tracer concentration as a function of time for various tundish arrangements after PCH Physico Chemical Hydrodynamics, 9, J. Szekely et al., "The Mathematical Modelling of Complex Fluid Flow Phenomena in Tundishes," 1987, Pergamon Journals Ltd.

improvement; here the tracer is being retained in the "pouring box" during an initial time period.

Finally in this series, the use of a slotted dam (Fig. 4.12) will provide for marked short circuiting; this behavior was to be anticipated from the velocity plots and is perhaps more clearly represented on the tracer isopleths.

Figure 4.13, also taken from Ref. [7], summarizes the results of the tracer dispersion runs. It is seen that in the absence of flow control we shall have the shortest breakthrough time. A dam and a weir arrangement may improve matters, and the use of two weirs and one dam will not really be beneficial. The slotted dam will provide only slight improvement over the results obtained in the absence of flow control.

Ultimately, one of the more important aspects of tundish performance is associated with the flotation of inclusion particles. Figure 4.14 shows the percentage of inclusion particle removal as a function of the particle diameter for the various tundish flow-control arrangements.

It is of interest to note that both ideal configurations, that is, perfect mixing and plug flow, tend to be better than any of the flow-control, configurations considered. The reason for this is quite simple: A significant dead volume is inherent in all the tundish flow arrangements. As expected, plug flow would represent the ideal flow-control arrangement, but this would be difficult to realize using conventional flow control.

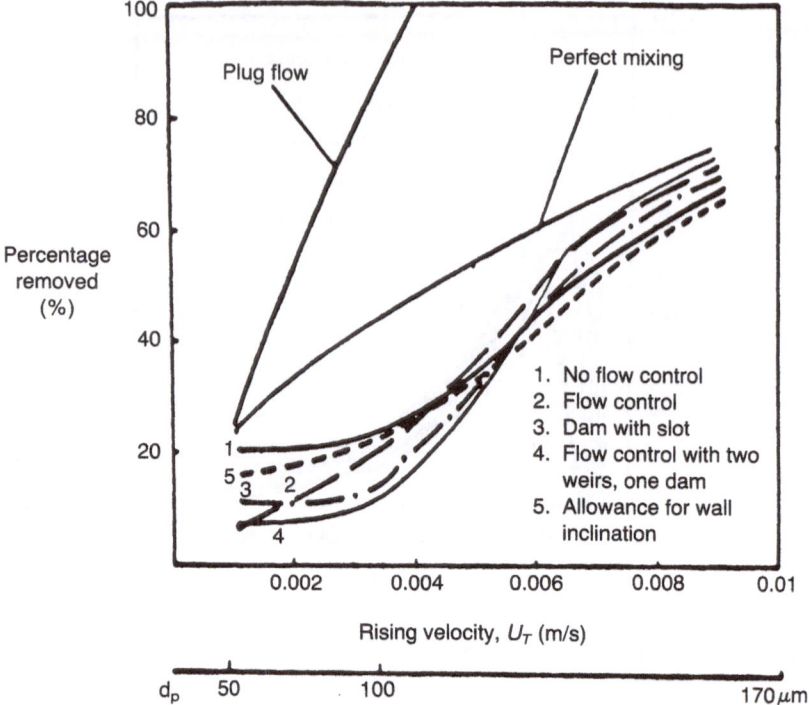

Fig. 4.14. Percentage of inclusion removal as a function of particle rising velocity and diameter.

It should also be remarked that the removal efficiency will depend on the particle size in a somewhat unexpected manner. With flow control, we can remove the larger particles with a good efficiency; however, in the absence of flow control we will retain the larger particles, but will be more efficient in removing the smaller ones. The behavior is readily explained by considering the fact that the more violent mixing in the absence of flow control should promote the upward movement of the smaller particles to an extent that could not be realized in a more controlled flow situation.

Up to the present time, there are very few plant scale measurements with which the predictions may be compared; a notable exception is provided by a recent paper jointly published by the engineers from Hoogovens and the present authors [8].

The tundish configuration is shown in Fig. 4.15 where it is seen that a particular design of slotted dam was being employed, having angled holes which provide an upward passage for the metal stream.

Figures 4.16 and 4.17 show a comparison between the experimentally measured and the theoretically predicted F curves, that is, the response of the system to a step change in the tracer concentration. The agreement between measurements and predictions appears quite good.

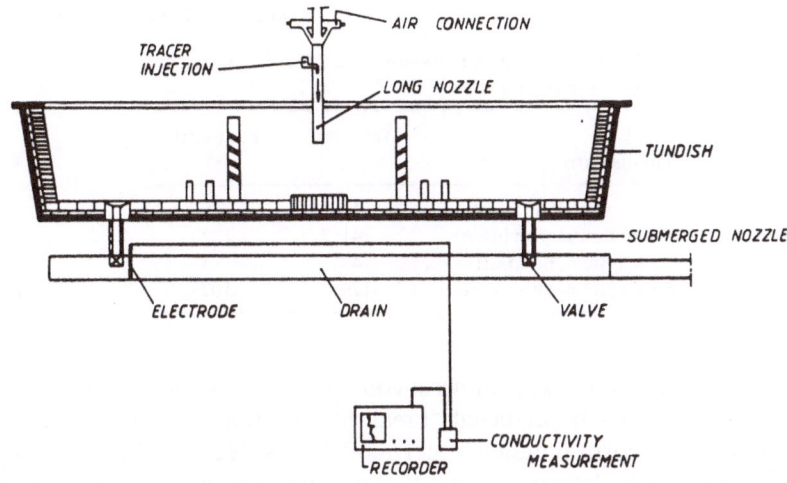

Fig. 4.15. Setup of physical experiments for Hoogoven's tundish system after Ilegbusi et al. [9].

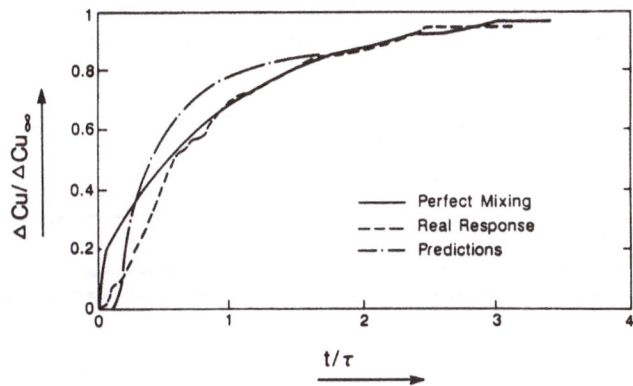

Fig. 4.16. Measured and predicted response of Hoogoven's steel system to a step change in tracer concentration in the absence of flow control after Ilegbusi et al. [9].

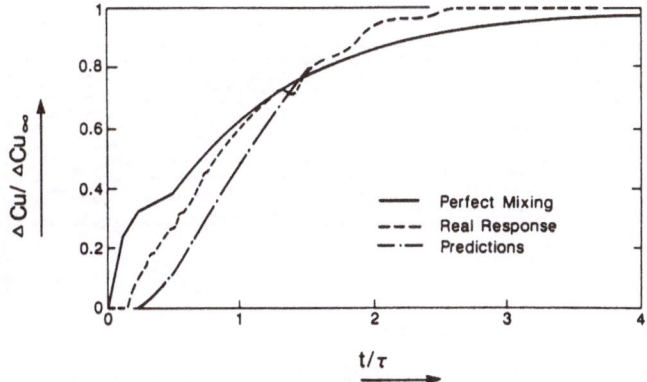

Fig. 4.17. Measured and predicted response of Hoogoven's steel system to a step change in tracer concentration in the presence of flow control after Ilegbusi et al. [9].

Table 4.3. Predicted and measured breakthrough times for Hoogoven's system

Tundish arrangement	Measured (s)	Predicted (s)
Water model, no insert	40	66
Water model with inserts	141	132
Steel system, no insert	49	73
Steel system with inserts	129	107

Table 4.3, taken from Ref. [9], shows a comparison between the experimentally measured and theoretically predicted "breakthrough times" for both the molten steel system and the corresponding water model tests. Here again the agreement seems quite reasonable throughout and it is of interest to note the measurements for the water model appear to differ quite significantly from the measurements obtained for molten steel.

This difference in behavior is no accident. Figures 4.18 and 4.19 show the computed velocity profiles at the center plane for the case of molten steel and for the

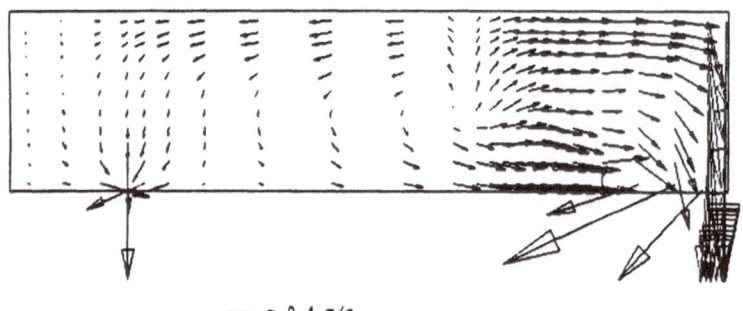

0.4 m/s

Fig. 4.18. Computed velocity field at the central plane of Hoogoven's water model in the absence of flow control after Ilegbusi et al. [9].

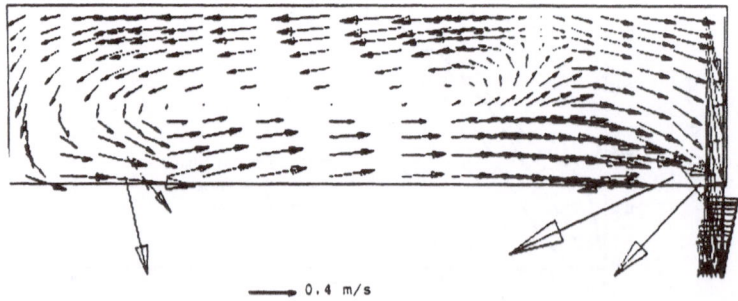

0.4 m/s

Fig. 4.19. Computed velocity field at the central plane of Hoogoven's steel system in the absence of flow control after Ilegbusi et al. [9].

corresponding water model, respectively. The quite significant differences in the flow patterns are readily apparent. The velocity field for the molten steel system is seen to "sweep over" the whole tundish, while the recirculation is more confined in the case of water system. This difference is remarkable, because the Froude numbers were the same and the Reynolds numbers were approximately equal. We should note, however, that we are dealing with a transitional flow, where the flow is highly turbulent in the inlet regions, and the turbulence decays as the fluid progresses toward the exit.

4.2.2 Discussion

The important conclusions that one may draw from these results are the following:

1) The velocity field, the pattern of turbulence, and the (somewhat idealized) behavior of inclusion particles are easily modeled for a three-dimensional system.
2) In the absence of flow control, significant short circuiting and by-passing will occur.
3) Short circuiting and by-passing may be significantly reduced by the use of flow-control arrangements; however, even for the optimal dam and weir geometries, a significant fraction of the tundish will act as a dead volume.
4) There is some evidence that water models will not fully replicate the behavior of molten steel systems, particularly in the absence of flow-control arrangements.

4.3 Conventional Tundishes: Heat Transfer

Typically, the molten steel stream enters the tundish at temperatures in the range of 1600°C and the inevitable heat losses will cause a temperature drop in the range of 10°–30°C. These heat losses themselves would not really constitute a major problem; however, it is a well-established fact that heat losses will also occur in the ladle, so that the temperature of the metal stream entering (and hence leaving) the tundish will vary with time. This is undesirable, because the structure and properties of the cast structures will, of course, depend on the superheat.

One of the interesting recent developments has been the use of tundish heaters, which enables one to supply thermal energy to the tundish itself, thus allowing a precise control of the steel temperature entering the mold of the continuous caster [10, 11].

As shown in Figs. 4.20 and 4.21, there are two basic possibilities for tundish heating:

1) the use of induction coils, and
2) the use of impinging plasma jet.

Mathematical modeling is indispensable in quantifying the behavior of these systems, because the usually deployed water models could not represent the high conductivity of molten steel (i.e., there would be a major disparity in the Prandtl numbers).

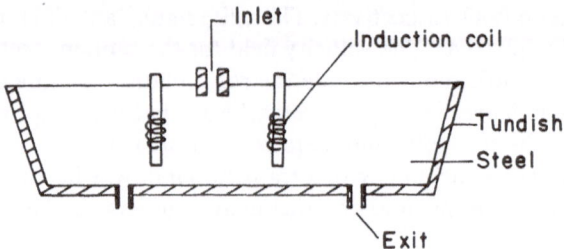

Fig. 4.20. Induction heating method in tundish.

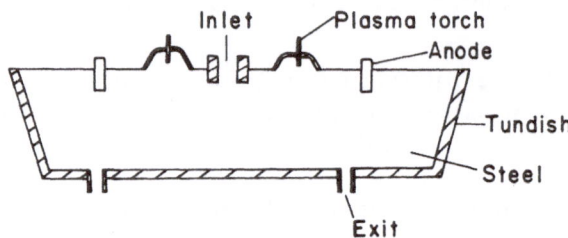

Fig. 4.21. Plasma heating method in tundish.

Table 4.4. Principal input parameters for system with heat transfer

Tundish length	6.79 m
Tundish width	0.65 m
Melt depth	0.75 m
Inlet stream velocity	7.67 m/s
Inlet nozzle diameter	54 mm
Outlet nozzle diameter	54 mm
Inlet stream temperature	1,600 K
Thermal conductivity of slag	3.4 m^2/s
Reference density	7,100 kg/m^3
Reference viscosity	0.001 m/kg·s
Specific heat	450.0 W/kg·°C

4.3.1 Computed Results

In the following, we shall present a selection of the computed results concerning the temperature profiles within the system and the rate of heat loss. The input parameters for these calculations are summarized in Table 4.4.

The principal results are given in Table 4.5, which shows the temperature of the exit stream for various tundish geometries and heating arrangements.

It is seen that in the absence of auxiliary heating, there is a significant temperature drop, which is somewhat larger in the absence of flow control.

When induction stirring is used, this tends to be more efficient and will benefit

Table 4.5. Exit temperatures for different tundish designs and auxiliary heating arrangements[a]

Tundish design	No auxiliary heating	Induction heating	Plasma heating
No flow control	1,852	1,858	1,867
Flow control	1,856	1,867	1,858

[a]Inlet temperature $T_0 = 1,873°K$; all values are in kelvins.

from flow control. In contrast, when the top metal surface is being heated by an impinging plasma jet, this will tend to be less efficient and would benefit somewhat from the absence of flow-control devices.

The following computed results will provide a better insight into the behavior of the system, and at the same time establish a technical basis for explaining these findings.

Behavior in the Absence of Auxiliary Heating

Figures 4.22 and 4.23 show selected isotherms in the absence and in the presence of flow-control arrangements without additional heating. A comparison of these plots clearly indicates that the metal stream cools more rapidly in the absence of flow control arrangements. When flow control is used in the form of dams and weirs,

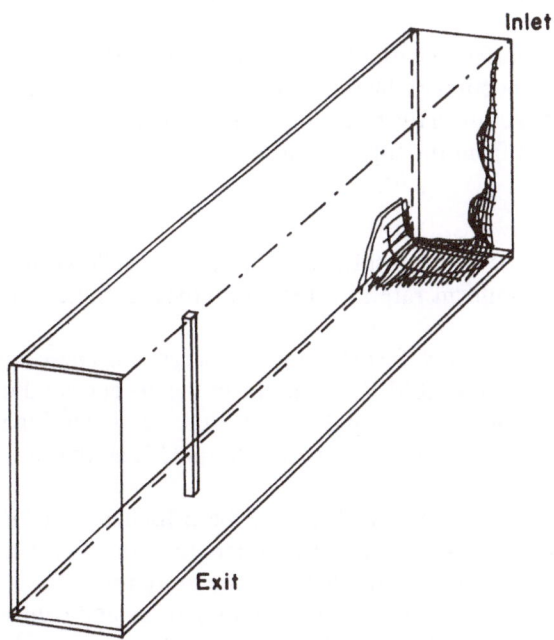

Fig. 4.22. Selected isotherm (1,865 K) profile in the absence of both auxiliary heating and flow control.

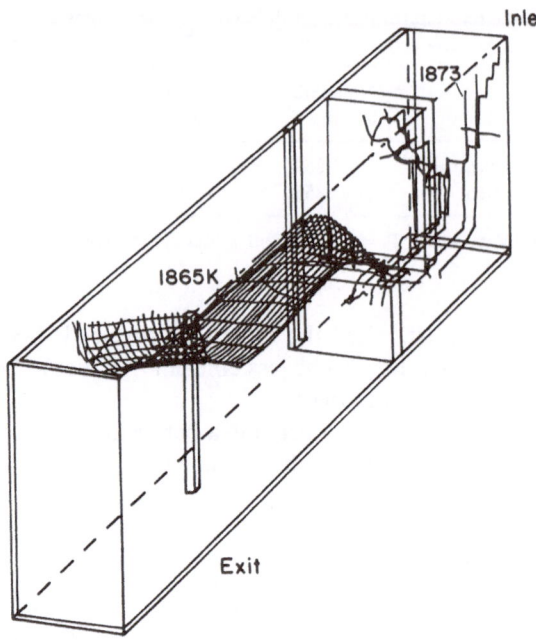

Fig. 4.23. Selected isotherms (1,865 K, 1,873 K) with flow control in the absence of auxiliary heating.

the 1,873 K isotherm is still discernible, and the 1,865 K extends well into the length of the tundish.

This behavior is consistent with physical reasoning; in the absence of flow control, there will be a larger portion of the tundish where the metal is well mixed, giving rise to a higher rate of heat loss in the near-entry region, owing to the strong nonlinearity of the radiation term from the top slag surface.

Induction Heating

In this arrangement, we shall consider that the metal, as it flows through the tundish, will be heated at a uniform rate at a location close to the exit from the dam, as indicated in Fig. 4.21.

Figures 4.24 and 4.25 show the behavior of this system. These figures respectively depict the 1,873 K and the 1,865 K isotherms in the absence and in the presence of flow-control arrangements. It is seen that the heat losses will be higher in the absence of flow control, as evidenced by the fact that the 1,873 K isotherm would be confined to the inlet region in this case.

In the absence of flow control, there will be a higher heat loss, and a greater spatial nonuniformity in the temperature. When flow control is being used, the heat loss is reduced and the temperature field will be more uniform.

Of course, the provision of the auxiliary heating will significantly increase the exit temperature from the tundish in any case, but this arrangement will be more effective in the presence of flow control.

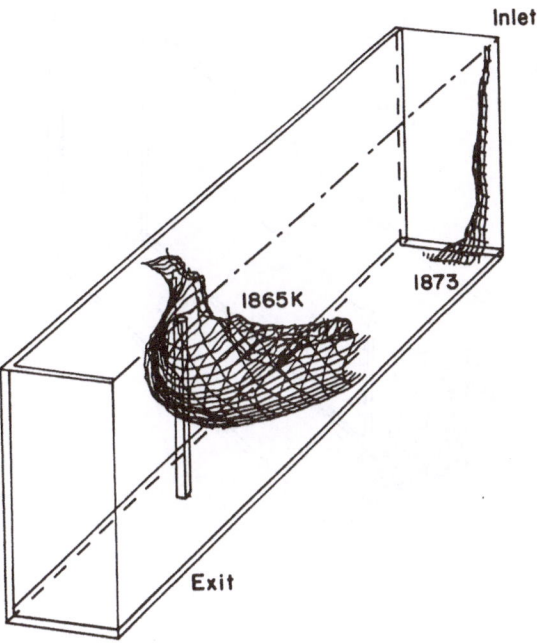

Fig. 4.24. Selected isotherms (1,865 K, 1,873 K) in the absence of flow control with induction heating.

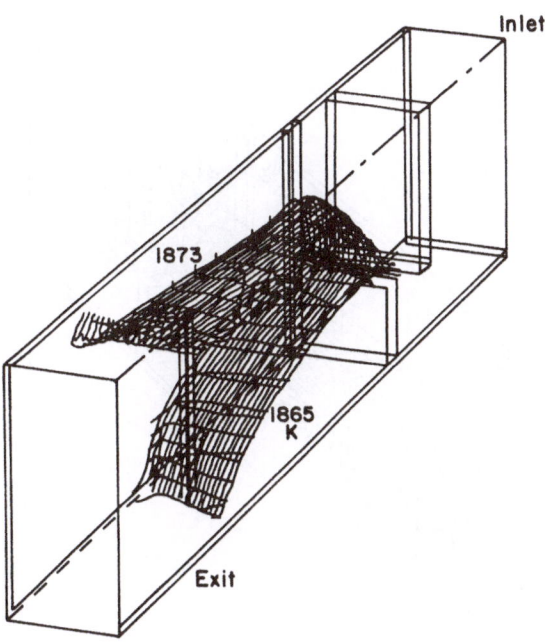

Fig. 4.25. Selected isotherms (1,865 K, 1,873 K) in the presence of flow control and induction heating.

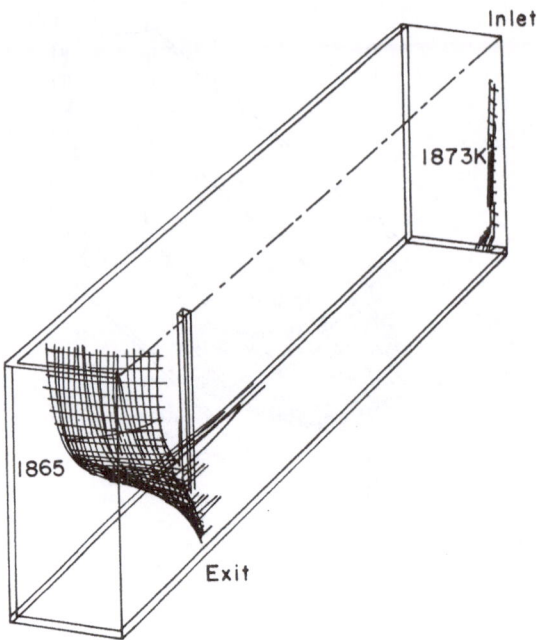

Fig. 4.26. Selected isotherms (1,865 K, 1,873 K) in the absence of flow control with plasma heating.

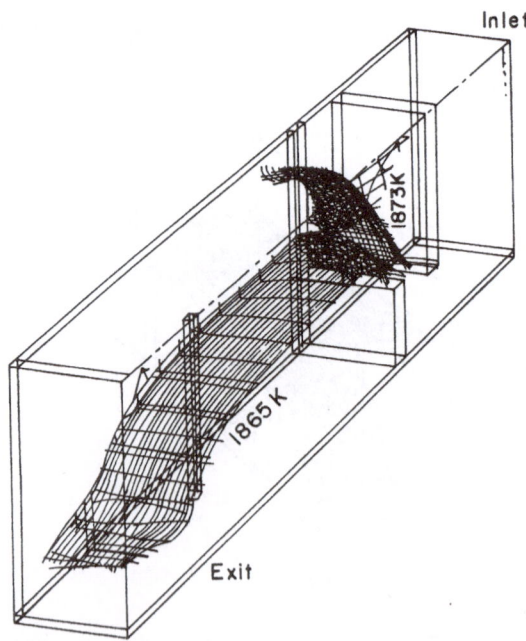

Fig. 4.27 Selected isotherms (1,865 K, 1,873 K) in the presence of flow control with plasma heating.

Plasma Heating of the Top Surface

Figures 4.26 and 4.27 show selected isotherms for plasma heating at the top surface. Inspection of these figures shows that for this case, the use of the flow control actually constitutes a drawback. The reason for this is clearly demonstrated in these plots. The quasistagnant area behind the dam does not participate in the heat transfer process, and the net result is that the 1,865 K isotherm is much closer to the exit plane than is the case without flow control.

This behavior again is quite consistent with physical reasoning; when the flow is stratified, heating the top surface is not an efficient way of adding thermal energy. In the absence of flow control, the strong mixing and high turbulence actually aid dispersion of the thermal energy supplied at the top. The provision of flow-control arrangements will reduce mixing and will thus interfere with the ready absorption of the thermal energy provided by the plasma jet.

4.3.2 Discussion

A mathematical representation has been developed to describe the temperature profile in tundishes, as affected by both flow-control and auxiliary heating arrangements.

The principal findings may be summarized as follows:

1) When no auxiliary heating is being provided, more significant heat losses will occur in the absence of flow-control arrangements. In any case, there will be a certain stratification of the melt as it approaches the exit regions. If multistrand casting arrangements were used, in either case, there would be significant differences in the temperatures of the different casting streams.
2) Auxiliary heating is a potentially attractive way of compensating for heat loss in the tundish and for providing a rather more precise temperature control in these systems.
3) In selecting auxiliary heating arrangements, however, care must be taken in both the choice of the heater and in its location. The point is that beyond the pouring box (i.e., outside the dam region, in the case of flow control), the melt is relatively quiescent and quasilaminar. This means that plasma heating will not be an effective way of providing thermal energy to an otherwise quiescent stream. In contrast, induction heating and the associated stirring would be an effective way of raising the tundish temperature in the presence of flow control.

The work presented here underlines the complexities inherent in tundish operation. Flow-control devices are needed in order to provide a quiescent melt and conditions conducive to the flotation of inclusions. At the same time, these conditions would be far from ideal for absorbing thermal energy supplied at the top free surface or for dispersion of alloying additions, for that matter.

One should note that even the treatment presented here was an oversimplification. In practice, the temperature of the ladle stream is neither steady nor constant, so that in ideal tundish operation one would wish to compensate for these factors. Indeed, the provision of such compensation should be the principal reason for using tundish heaters.

It is hoped that the computational techniques presented could provide the technical basis for the improved operation of tundish heating.

4.4 Magnetic Flow Control in Tundishes

One of the interesting new ideas that has emerged very recently is the use of magnetic fields to control and streamline the flow in tundishes. The innate principle of magnetic flow control is quite simple. If a moving fluid is introduced into a static magnetic field, there will be an electromagnetic "braking force" exerted on the fluid, which is defined as

$$\mathbf{F} = \mathbf{J} \times \mathbf{B}, \tag{4.14}$$

in which $\mathbf{B}$ is the magnetic field and $\mathbf{J}$ is the current density defined from Maxwell's relations as

$$\mathbf{J} = \sigma_e(\mathbf{E} + \mathbf{v} \times \mathbf{B}), \tag{4.15}$$

where σ_e is the electical conductivity and $\mathbf{E}$ is the electric field.

The forms of Eqs. (4.14) and (4.15) clearly suggest that the higher the melt velocity, the greater will be the braking force, so that a properly oriented field will tend to even out the velocities and minimize the velocity gradients. By the same token, a properly oriented field will tend to reduce the vorticity. A good basic discussion of these fundamental issues is available in the text by Shercliffe [12].

Figure 4.28 and Table 4.6 summarize the different cases including the field arrangements that have been considered here. The absence of field will represent the base case; then we can examine the effect of relatively narrow field (magnetic gate), which would be oriented perpendicular to the principal flow direction. The other alternative is to impose such a field across the whole tundish. Clearly, the latter solution would tend to be more expensive.

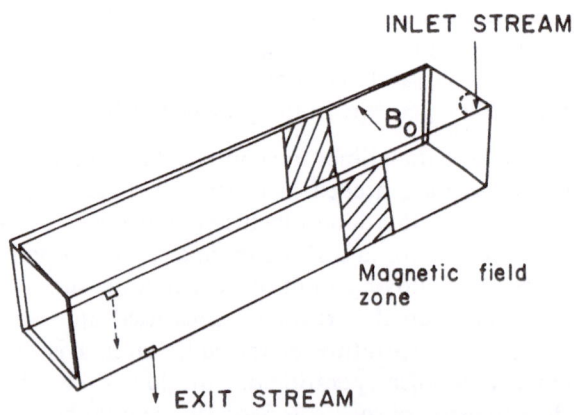

INLET STREAM

B_0

Magnetic field zone

EXIT STREAM

Fig. 4.28. Schematic sketch of a typical tundish with magnetic field B_0 applied in the lateral direction.

Table 4.6. Tundish arrangements investigated for the magnetic-flow-control calculations

Case	Tundish arrangement
a	No magnetic field
b	Magnetic field of 3 kg imposed on a section 70 cm from inlet in the lateral direction
c	Magnetic field of 3 kg imposed across the whole tundish in the lateral direction

This problem is readily formulated by including an electromagnetic body force in the fluid flow equations, such that the latter take the general form

$$\rho U \cdot \nabla U = -\nabla p + \nabla \cdot (\mu_{\text{eff}} \nabla U) + \mathbf{F}, \tag{4.16}$$

where $\mathbf{F}$ (F_x, F_y, F_z) is the electromagnetic force per unit volume given in Eq. (4.15). This term represents a coupling between the velocity field within the conducting medium and the externally imposed magnetic field. This force has to be obtained through the solution of Maxwell's equations.

4.4.1 Computed Results

In the following, we shall present a selection of the computed results for tundishes used in conventional continuous casting systems. The input parameters are similar to those already given in Table 4.2.

Figures 4.29a and 4.29b show the computed velocity fields in two longitudinal slices of the tundish in the absence of an external field (case a). Here Fig. 4.29a corresponds to the central plane of symmetry, while Fig. 4.29b depicts a plane close to the side wall. These plots reproduce a behavior that has been seen in previous

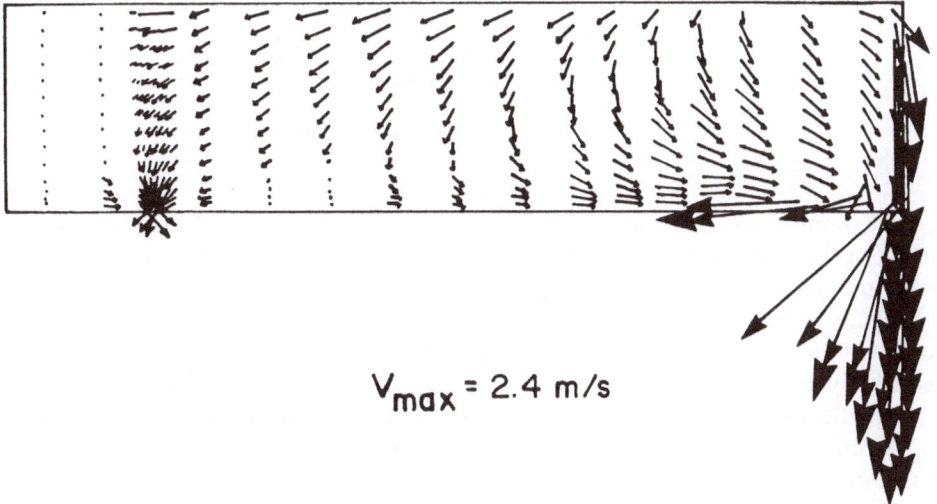

$$V_{\text{max}} = 2.4 \text{ m/s}$$

Fig. 4.29a. Velocity field in the vertical plane near the center (0.15y) for no magnetic field, case a.

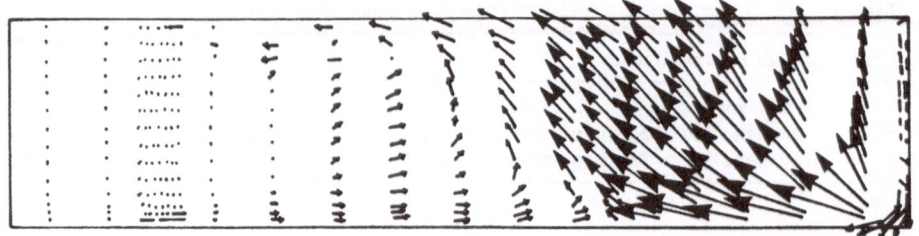

$$V_{max} = 1.0 m/s$$

Fig. 4.29b. Velocity field in the vertical plane near the side wall ($0.8y$) for no magnetic field, case a.

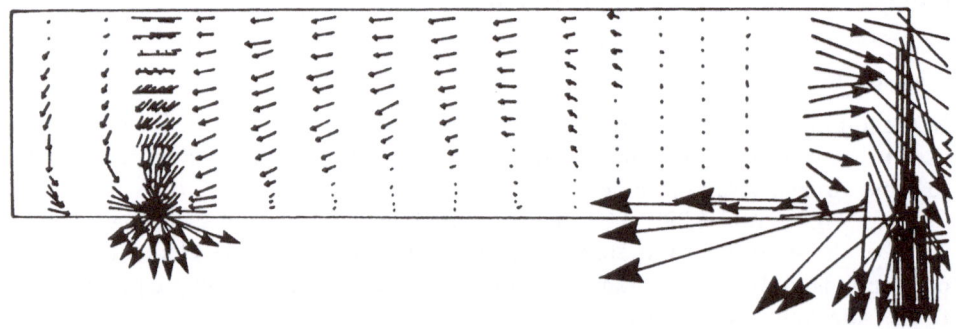

$$V_{max} = 2.4 \ m/s$$

Fig. 4.30a. Velocity field at $0.15y$ with $\mathbf{B}_0 = 3$ kG, case b.

publications: a strong recirculating zone in the vicinity of the inlet and a marked effect of the exit nozzle in the central plane. A marked upward flow near the side wall in the entrance and a relatively quiescent zone near the exit are also observed.

Figure 4.30 shows the corresponding plots for case b, that is, a 3 KG field imposed on a section 70 cm distance from the inlet. The very marked difference in behavior is readily apparent. The flow is "straightened out" upon passing through the "magnetic gate," and a very uniform flow field is being established. The flow uniformities will, of course, persist in the exit region, because there is no damping force there.

Figure 4.31 shows the corresponding plots, but for case c, when a 3 KG field is imposed across the whole of the tundish.

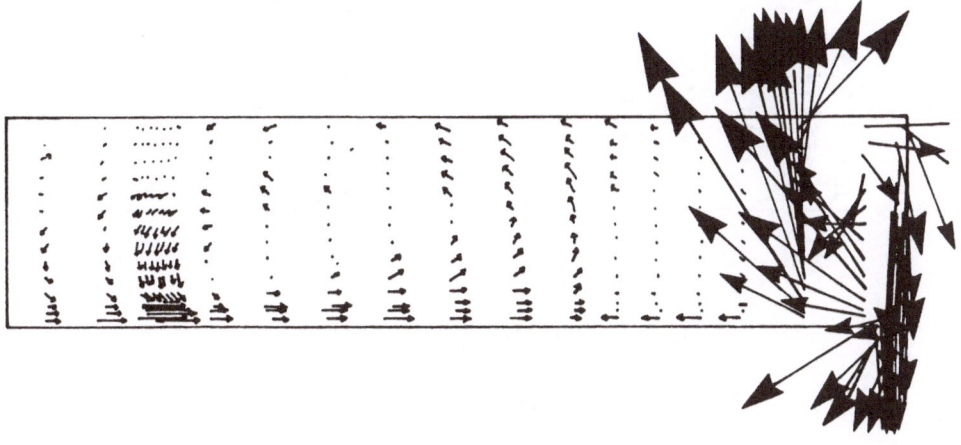

$$V_{max} = 1.0 \text{ m/s}$$

Fig. 4.30b. Velocity field at $0.8y$ with $\mathbf{B_0} = 3$ kG, case b.

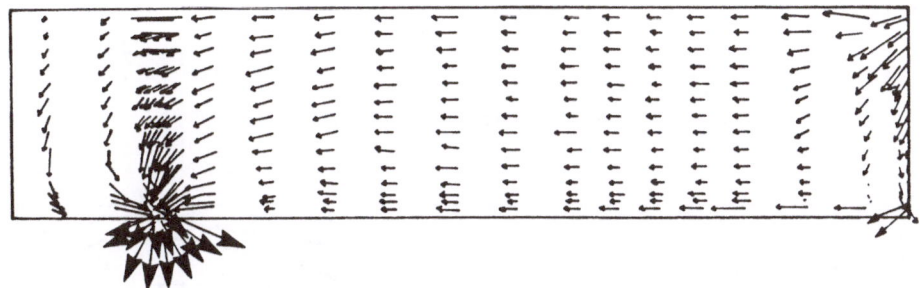

$$V_{max} = 1.9 \text{ m/s}$$

Fig. 4.31a. Velocity field at $0.15y$ with $\mathbf{B_0} = 3$ kG, case c.

A comparison of Figs. 4.31a and 4.31b with the corresponding plots given in Figs. 4.29 and 4.30 shows a dramatic effect. It is seen that the very marked recirculating flow field is effectively killed off in the entrance region, resulting in a highly uniform velocity field throughout the tundish, with the exception of the exit nozzle.

This startling behavior is readily explained by the fact that the field imposed in the x direction will tend to damp the vorticity in the x-z plane, and, hence, will deflect the U_z velocity component. This behavior will have a major practical significance, in that any inclusion particles that are formed will not be carried to the bottom of the tundish, but rather should have a much shorter distance to travel.

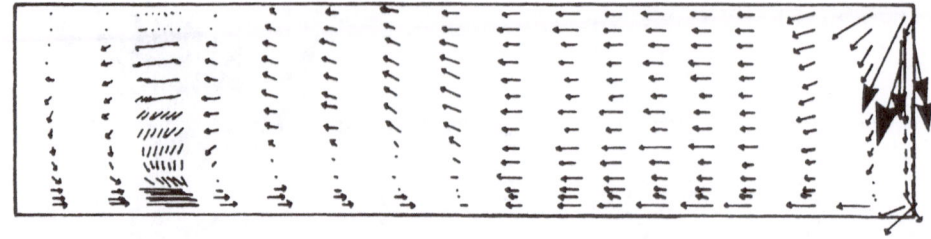

$$V_{max} = 0.35\,m/s$$

Fig. 4.31b. Velocity field at $0.8y$ with $\mathbf{B_0} = 3$ kG, case c.

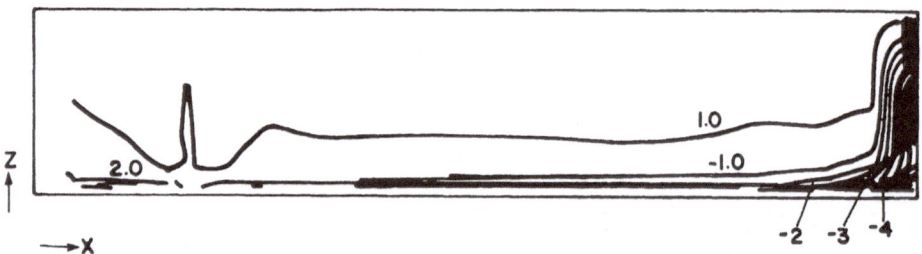

Fig. 4.32a. Vorticity at $0.15y$ with no magnetic field, case a.

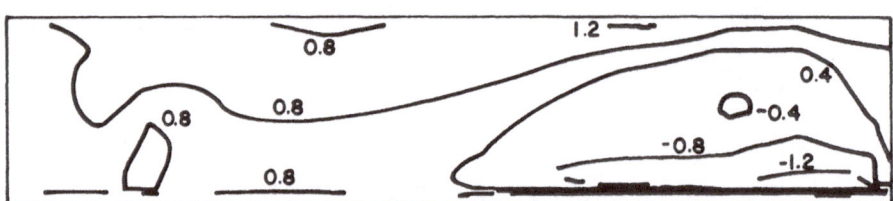

Fig. 4.32b. Vorticity at $0.8y$ with no magnetic field, case a.

Of course, the uniform field imposed throughout the tundish cannot damp the vertical outflow of the metal, because this is an exit condition imposed on the system; but closer inspection of Fig. 4.31 shows that the vertical velocity component is indeed being damped, once we move away from the exit.

From the standpoint of fundamental fluid flow considerations, it is of interest to examine the behavior of cases a and c on the vorticity plots. This is done in Figs. 4.32 and 4.33. It is readily seen that in the absence of magnetic flow control, the vorticity will be quite high throughout the system, while magnetic flow control will be very effective in minimizing the vorticity throughout the tundish domain. This

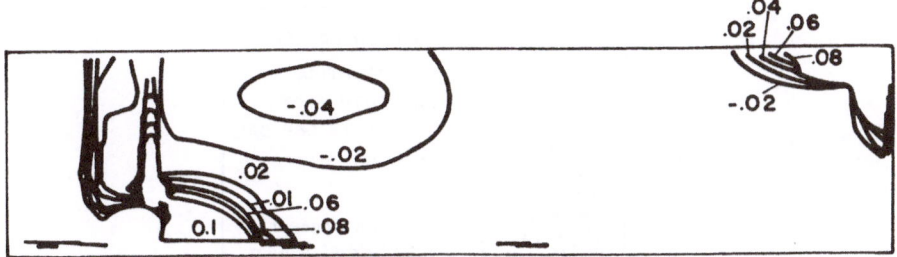

Fig. 4.33a. Vorticity at $0.15y$ with $B_0 = 3$ kG, case c.

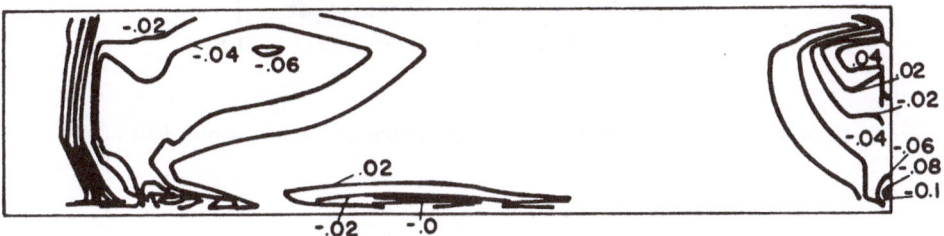

Fig. 4.33b. Vorticity at $0.8y$ with $B_0 = 3$ kG, case c.

behavior is of interest, because it is consistent with the qualitative arguments that have been advanced in the introductory section. Another important point to be made is that by reducing the vorticity in the vicinity of the exit nozzle, we shall also diminish the tendency for vortex formation.

4.4.2 Tracer Dispersion

In Section 4.2 we examined the effect of various tundish flow control arrangements on the tracer dispersion curves. This plot is reproduced in Fig. 4.34 with the addition of the curve produced by the use of a magnetic gate, corresponding to case b. The dramatic effect is readily seen; the breakthrough time is markedly lengthened, and, now, the tracer dispersion is close to being symmetric about the nominal holding time in the vessel. These results would seem to indicate a relatively close approach to plug flow.

4.4.3 Inclusion Removal

One of the key purposes of employing deep and large tundishes is that these will facilitate the efficient removal of inclusion particles.

Figure 4.35 shows a plot of previously calculated results concerning inclusion removal, for various flow control arrangements and also for the idealized plug flow and completely mixed vessels. This plot also shows the computed behavior for case b. It is seen that the use of a magnetic gate offers a very marked improvement in the system performance over all the other previously computed cases.

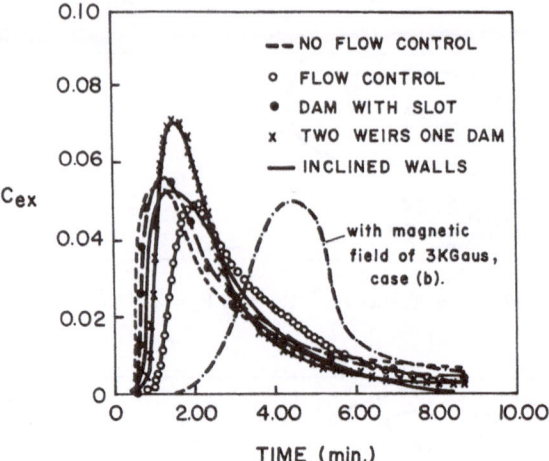

Fig. 4.34. Tracer concentration at the exit against time for various tundish designs.

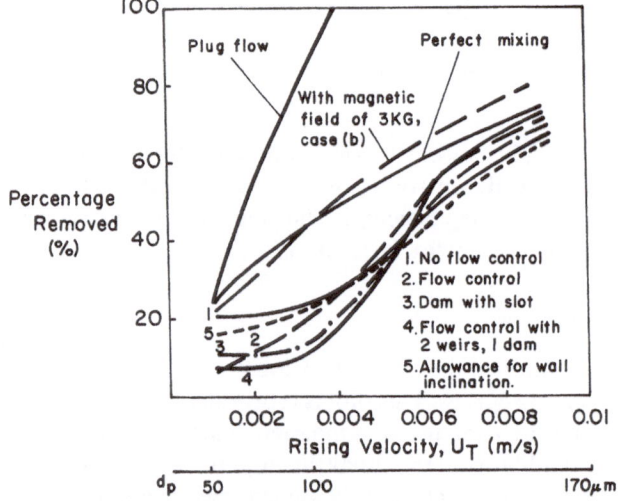

Fig. 4.35. Calculated inclusion removal rates for various tundish designs.

4.4.4 Discussion

The computed results show that the imposition of magnetic field has a dramatic and potentially beneficial effect on the transport phenomena within the tundish. The principal findings in this regard may be summarized as follows:

1) If a magnetic gate is imposed on the tundish, that is, a field over a limited tundish length, having an orientation perpendicular to the main direction of flow, this will have the effect of "straightening out" the flow, and providing a much more uniform, nearly parallel velocity distribution, downstream of the gate.

2) Such an arrangement as previously described will cause the flow to approach plug flow conditions, with a corresponding dramatic increase in the breakthrough time and in the efficiency of inclusion removal.
3) An even more dramatic effect can be achieved by imposing a uniform field throughout the tundish. Under these conditions, the field will virtually eliminate the strong recirculating zone near the inlet and will provide for a nearly uniform, parallel flow throughout the tundish. Conditions of this kind would be particularly conducive to the effective flotation of inclusion particles.

At this point, it is worthwhile to reemphasize the scientific principles involved. When a magnetic field is being imposed on a moving, conducting fluid, the coupling between them will give rise to an electromagnetic force that will seek to oppose the fluid motion in a strongly directional way. More precisely, the field will seek to suppress the vorticity (defined in terms of velocity gradients) exponentially within the plane of the imposed field. The net result of this is the reduction of velocity gradients and the tendency toward a parallel flow arrangement. This is the optimally desired configuration for tundishes.

Another important effect of the field is that through minimizing the vorticity the imposition of the field will also tend to reduce the tendency for vortex formation, which can be a serious problem, particularly when emptying tundishes.

Finally, a comment should be made on two other potential applications of magnetic fields, in conjuction with tundish design and operation.

One of these is the tendency of a stationary field to oppose instabilities and, hence, the formation of surface waves. It is an established fact that surface waves may form in tundishes when either the inflow or the outflow rate is changed. The presence of a field will tend to minimize the amplitude of the waves.

Perhaps an even more potential application of magnetic field, which will be the subject of the next section, is their use in the shallow, smaller tundishes needed in nonconventional continuous casting processes. Here, it is crucial to minimize the recirculating flows and the disturbance created by the inlet and the exit streams, especially since these will not be damped out readily for the shallow melt depths and short retention times.

While the experimental verification of these modeling results is clearly desirable, it is thought that the equations used for representing the physical processes considered here, that is, Maxwell's equations for the magnetic field effects and the Navier–Stokes equations for the velocity fields, are well enough established, as are the computational techniques used, that the results can be viewed with a good degree of confidence.

4.5 Shallow Tundish Behavior

All the preceding considerations apply to conventional tundishes, which are typically some 2–7 m long, about 0.5–1 m wide, and some 0.8–1 m deep, with nominal metal residence times in the region of 4–20 min.

However, there is emerging a new class of tundishes in conjuction with novel

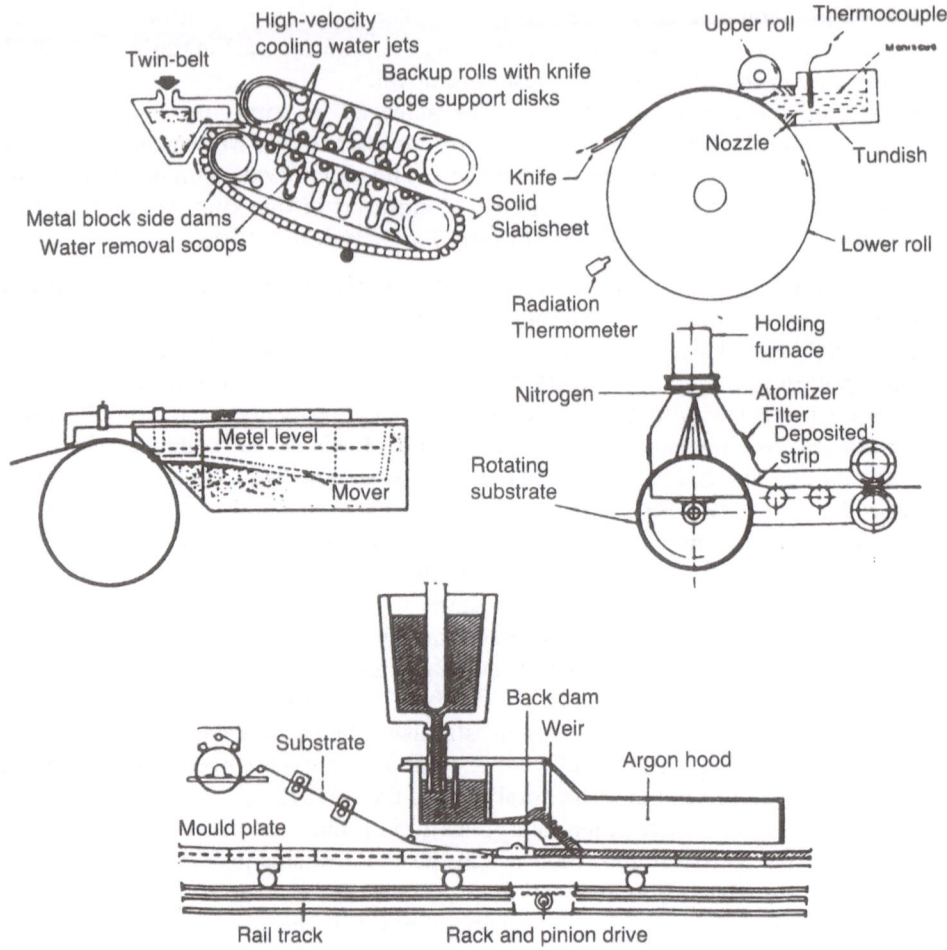

Fig. 4.36. Schematic sketch of some new continuous casting systems [19].

continuous casting techniques, which are rather different in shape from their conventional counterparts.

A good discussion of some of these new continuous casting systems is available in two recent symposium proceedings [13, 14]. Figure 4.36, taken from an article by Birat [13], shows a schematic sketch of these new developments. In short, we may consider single-roll or twin-roll systems, belt casters, and the SMS system, which involves an unconventional, tapered mold.

A common feature of all these systems is that the molten metal stream entering the solidification zones (the rolls, belt, or mold) has to pass through a buffer vessel or a tundish. The actual size (and often the shape) of these tundishes tends to be proprietary, but as a reasonable estimate, one may consider these to be about 1 m long, about 1 m wide and relatively shallow, with a depth of about 0.25 m. It is to be noted that somewhat similar tundishes are being used also in a number of

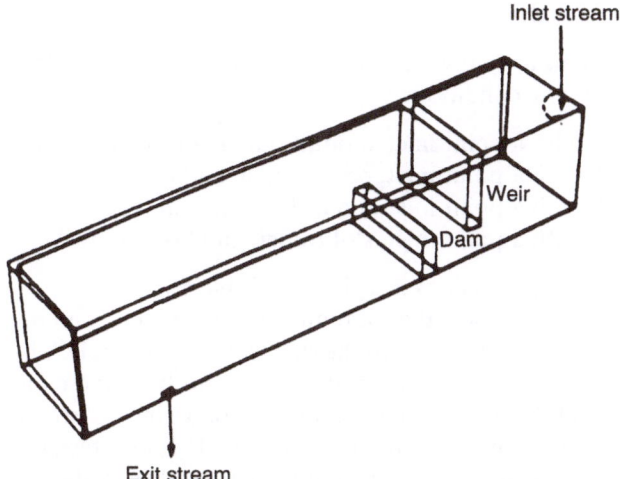

Fig. 4.37a. Typical tundish arrangement with point inlet and point exit [19].

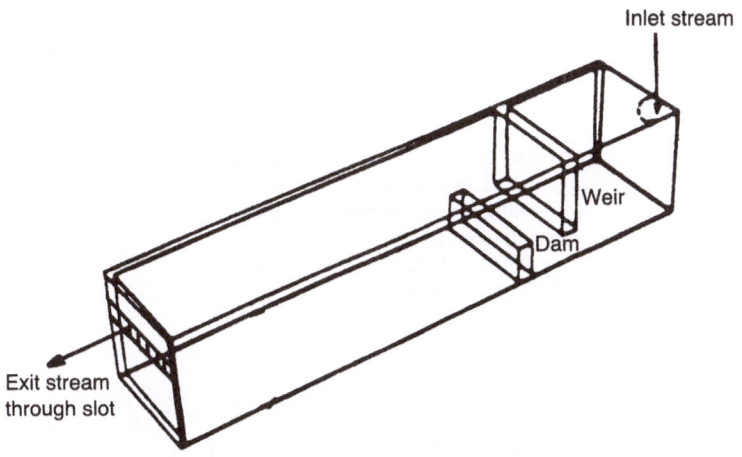

Fig. 4.37b. Typical tundish arrangement with point inlet and slot exit [19].

nonferrous solidification processing applications, notably in the continuous casting of aluminum, copper, and other alloys.

Up to the present time, relatively little attention has been paid to flow phenomena in these systems. In view of the analogy to conventional continuous casting systems, it is quite likely that flows in these systems could play a key role in the ultimate performance of the system, and that the judicious use of flow-control devices could mean the difference between success and failure in many applications.

The purpose of this section is to present some preliminary calculations on the behavior of somewhat idealized tundish systems. The main objective is to provide insight into the general flow behavior in these systems and to illustrate the role that system geometry and flow control may play.

4.5.1 Computed Results

Two idealized systems are shown in Figs. 4.37a and 4.37b (p. 87). We shall present results for the following conditions:

1) shallow tundish with a point inlet, a point outlet, and no flow control;
2) shallow tundish with a point inlet, point outlet, and flow control;
3) shallow tundish with a point inlet and a slot outlet, and no flow control; and
4) shallow tundish with a point inlet, slot outlet, and flow control.

The principal input parameters are given in Table 4.7.

Figures 4.38a and 4.38b show the computed flow field for a point inlet and a point outlet in the central plane and close to the side walls, respectively. The very marked recirculation and short circuiting are readily apparent from these plots.

Figures 4.39a and 4.39b show corresponding plots in the presence of a flow-control arrangement using one dam and one weir. It is of interest to note that in some contrast to the behavior of conventional tundishes, a dam and a weir do modify the flow field, but to a very insignificant extent.

Table 4.7. Principal input parameters for shallow tundish systems

Tundish length	1 m
Tundish width	0.8 m
Tundish height	0.25 m
Inlet nozzle diameter	4 cm
Outlet nozzle diameter (point outlet)	4 cm
Outlet slot height (slot outlet)	3 cm

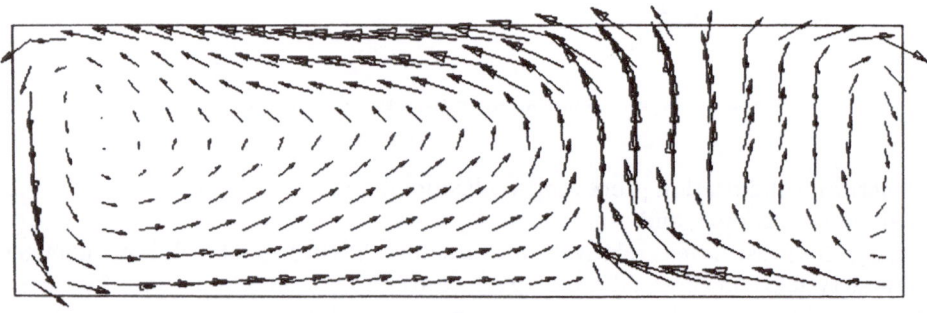

$\longrightarrow$ 5 cm/s

Fig. 4.38a. Flow field near central plane (0.1y) for a point inlet and point exit in the absence of flow control [19].

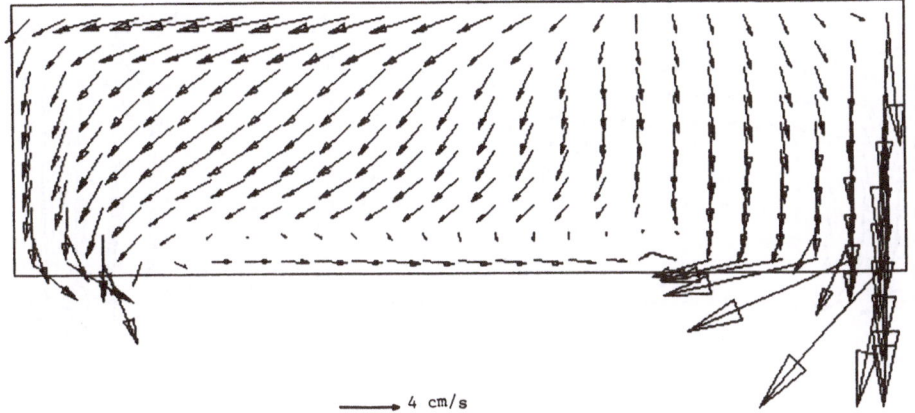

→ 4 cm/s

Fig. 4.38b. Flow field near side wall (0.8*y*) for a point inlet and point exit in the absence of flow control [19].

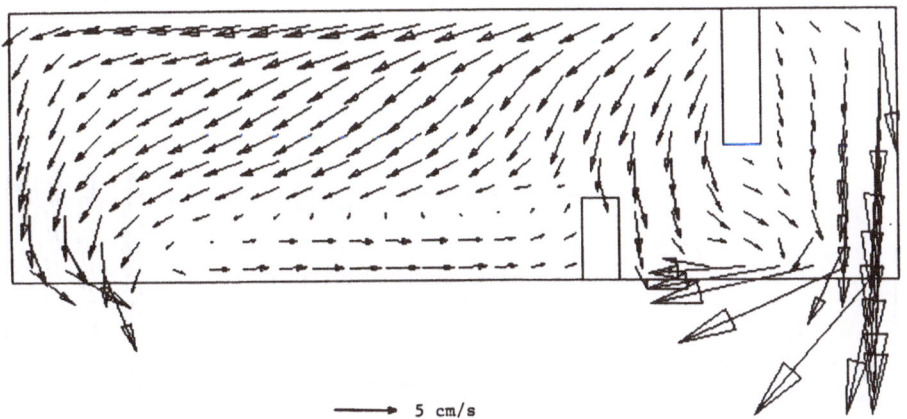

→ 5 cm/s

Fig. 4.39a. Flow field near central plane (0.1*y*) for a point inlet and point exit with one dam and one weir [19].

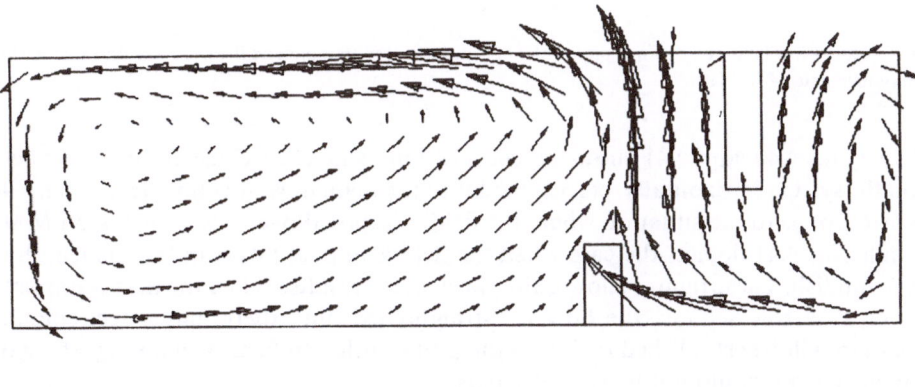

→ 4 cm/s

Fig. 4.39b. Flow field near side wall (0.8*y*) for a point inlet and point exit with one dam and one weir [19]

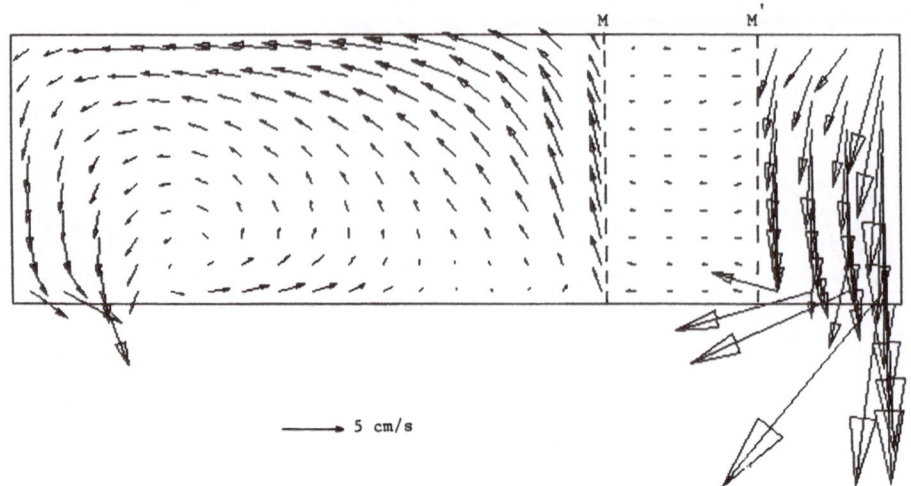

Fig. 4.40a. Flow field at 0.1y for a point inlet and point exit with magnetic field of 1 kG imposed within magnetic gate MM' [19].

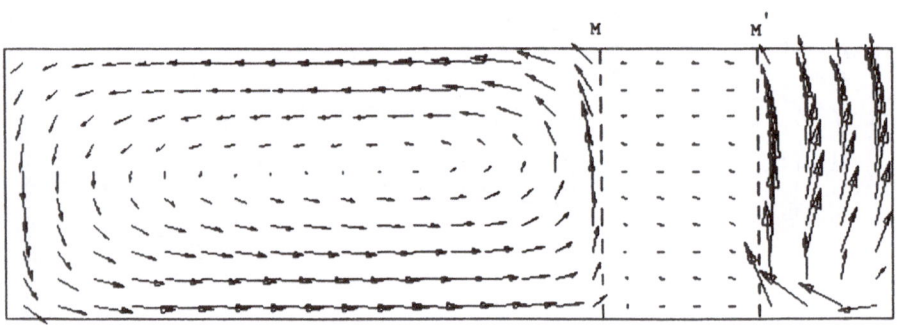

Fig. 4.40b. Flow field at 0.8y for a point inlet and point exit with magnetic field of 1 kG imposed within magnetic gate MM' [19].

Figures 4.40. and 4.41 show the effect of imposing a 1-kG electromagnetic field on the system horizontally, perpendicular to the main flow direction. Here, Fig. 4.40 corresponds to the situation where the field is imposed over a narrow region MM', while Fig. 4.41 depicts the case when the field is imposed over the whole length of the tundish. These figures show quite spectacular results. When the field is imposed over a narrow section, the flow is "straightened out" there, but a recirculating pattern will be established owing to the point outlet. In fact, such a magnetic gate arrangement would not be very effective.

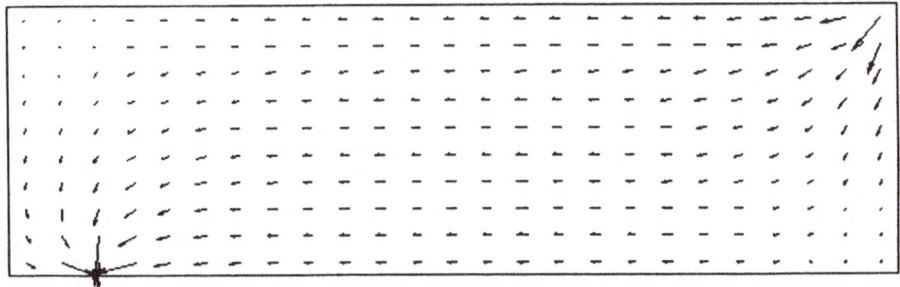

16 cm/s

Fig. 4.41a. Flow field at 0.1*y* for a point inlet and point exit with magnetic field of 1 kG imposed on whole tundish [19].

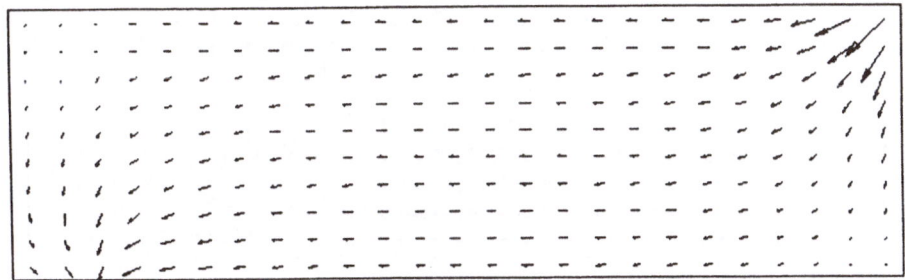

5 cm/s

Fig. 4.41b. Flow field at 0.8*y* for a point inlet and point exit with magnetic field of 1 kG imposed on whole tundish [19].

In contrast, as seen in Fig. 4.41, the imposition of a field along the whole length of the tundish will provide an excellent approximation to the classical plug flow arrangement.

Calculations have been carried out with lesser fields and it was found that one would need a field strength of 0.5–1 kG to have a significant effect on the behavior of the system.

Figures 4.42a and 4.42b show the behavior of the system with a point inlet and a horizontal slot outlet in the absence of flow control arrangements. This will be the

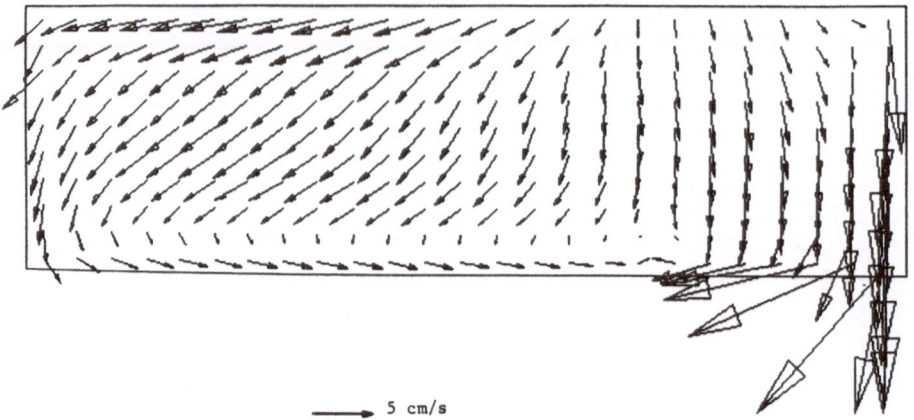

5 cm/s

Fig. 4.42a. Flow field near central plane (0.1y) for a point inlet and slot exit in the absence of flow control [19].

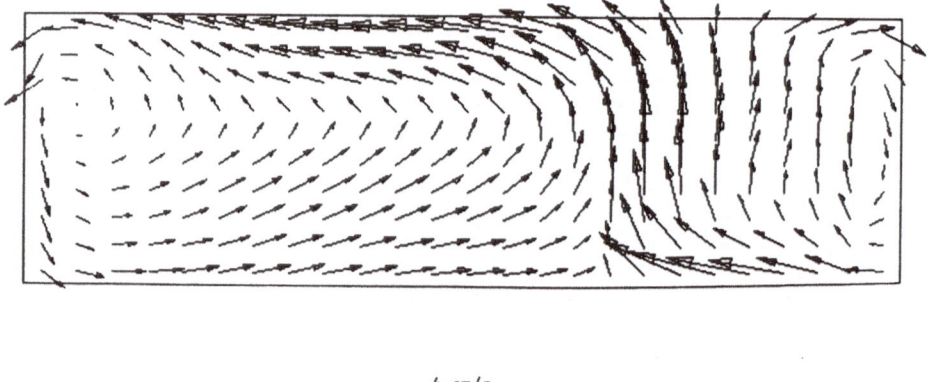

4 cm/s

Fig. 4.42b. Flow field near side wall (0.8y) for a point inlet and slot exit in the absence of flow control [19].

case corresponding to the majority of single-roll casting systems. Inspection of these plots is instructive, because they show marked recirculation and quite significant differences in the flow field in the exit plane.

Figure 4.43 depicts the effect of placing a dam and weir in the tundish in an attempt to control the flow. It is seen that these flow-control arrangements will modify the behavior of the system in the vicinity of the inlet region, but the bulk and the exit regions are not markedly affected.

Figures 4.44 and 4.45 illustrate the effect of imposing a 1-kG magnetic field on the system close to the inlet region and on the whole tundish, respectively. It is seen that these will have a marked, highly desirable effect on the structure of the flow. As expected, the magnetic field imposed on the whole tundish will play a major role in straightening out the flow and providing a very uniform flow field.

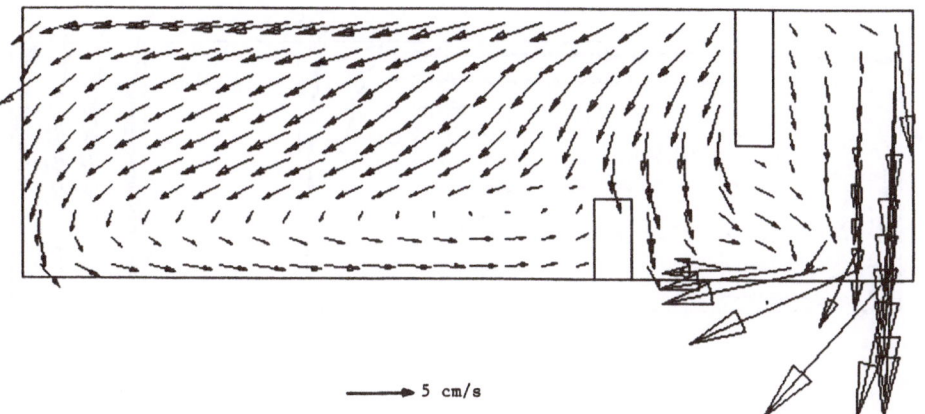

----→ 5 cm/s

Fig. 4.43a. Flow field near central plane (0.1y) for a point inlet and slot exit with one dam and one weir [19].

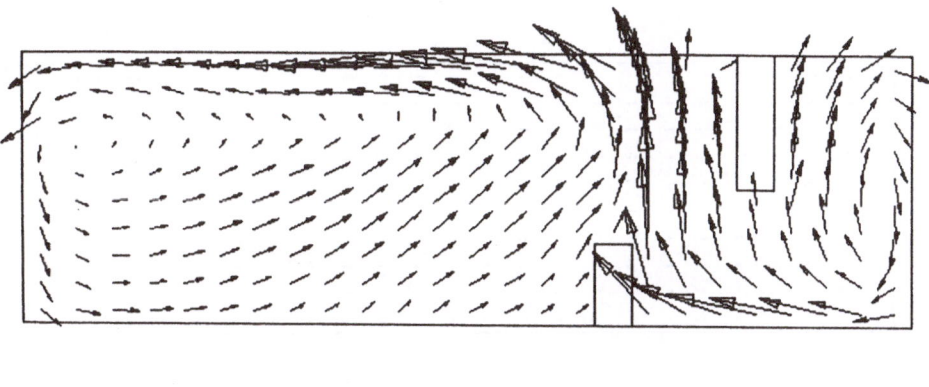

----→4 cm/s

Fig. 4.43b. Flow field near side wall (0.8y) for a point inlet and slot exit with one dam and one weir [19].

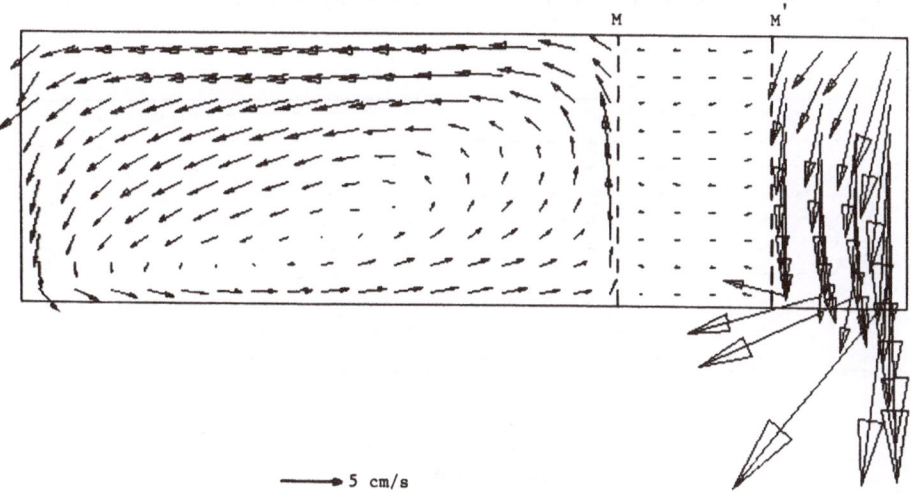

----→ 5 cm/s

Fig. 4.44a. Flow field at 0.1y for a point inlet and slot exit with magnetic field of 1 kG imposed within magnetic gate MM' [19].

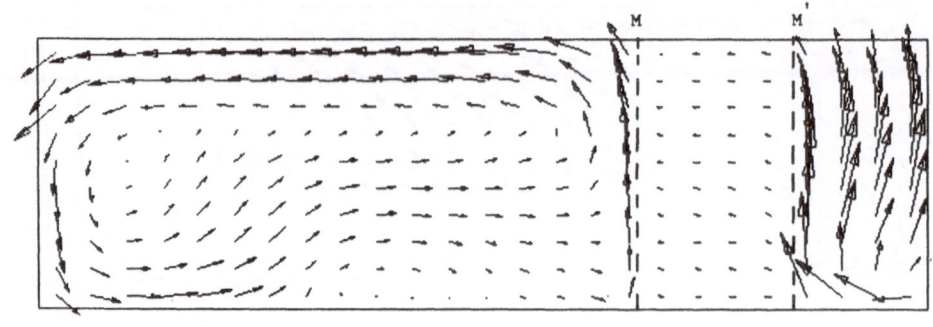

$\longrightarrow$ 4 cm/s

Fig. 4.44b. Flow field at $0.8y$ for a point inlet and slot exit with magnetic field of 1 kG imposed within magnetic gate MM' [19].

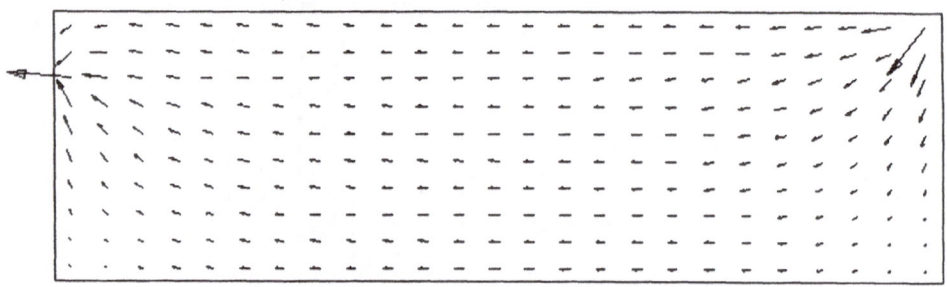

$\longrightarrow$ 16 cm/s

Fig. 4.45a. Flow field at $0.1y$ for a point inlet and slot exit with magnetic field of 1 kG imposed on whole tundish [19].

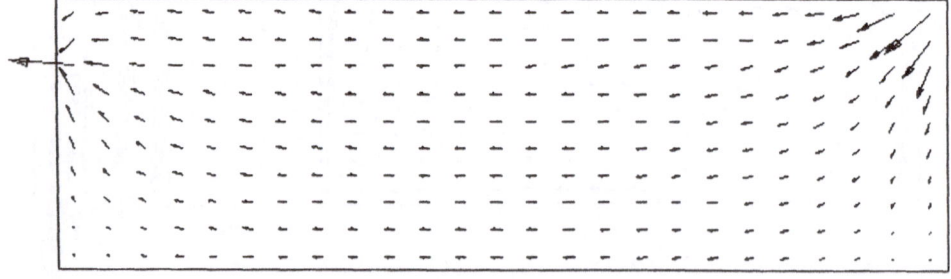

$\longrightarrow$ 5 cm/s

Fig. 4.45b. Flow field at $0.8y$ for a point inlet and slot exit with magnetic field of 1 kG imposed on whole tundish [19].

One key concern in the operation of tundishes, especially shallow ones with a point outlet, is whether vortexing will occur. This is an important consideration because vortexing, which is particularly prevalent for shallow melt depths, may result in flow instabilities, poor surface quality, and gas entrainment. The tendency for vortex formation is readily assessed by computing the major vorticity component in the vicinity of the exit nozzle.

Table 4.8 shows the computed maximum absolute values of the vorticity. It is seen that, as expected, the vorticity will be quite high in the case of a point exit. The value of the vorticity, and, hence, the tendency for vortex formation, may be reduced by having flow-control devices, and in an even more pronounced manner by imposing a magnetic field. As expected, the higher the field, the larger will be its effect.

Figure 4.46 shows a set of tracer dispersion curves for a point inlet and a point exit. It is seen that very significant short circuiting will occur, both in the absence

Table 4.8. Maximum absolute values of major vorticity component at exit of shallow tundishes

Tundish type	Point exit	Slot exit
No flow control	0.32	0.1
Flow control (one weir, one dam)	0.28	0.073
Magnetic flow control $(B_0 = 0.1 \text{ kG})$	0.22	0.036
Magnetic flow control $(B_0 = 1 \text{ kG})$	0.13	0.013

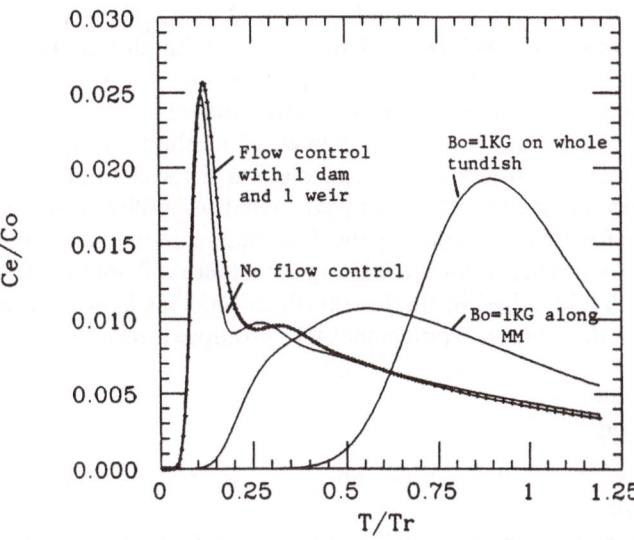

Fig. 4.46. Tracer dispersion curves for the tundish with point inlet and point exit [19].

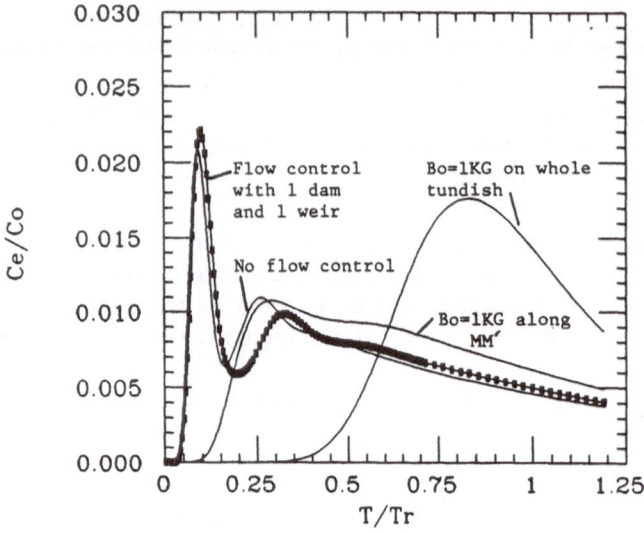

Fig. 4.47. Trace dispersion curves for the tundish with point inlet and slot exit [19].

of flow control and when a dam and a weir are being used. This is the expected behavior, because such systems represent a textbook example of flow maldistribution and short circuiting. It is seen on this plot, however, that the use of a magnetic field of 1 kG across the whole tundish length will have a significant effect in ameliorating the situation, indeed providing quite a good approximation to ideal plug flow. The use of a magnetic gate, that is, a field over a limited tundish length would have a rather less, but still highly beneficial, marked effect.

Figure 4.47 shows the behavior of the system with slot outlet, which would correspond to roll casting. It is seen that here, again, very significant by-passing would occur both in the absence of flow control and when conventional dams and weirs are deployed in the vicinity of the inlet. Here, the imposition of a magnetic field has an even more dramatic effect in making the system approach plug flow conditions. Even relatively low fields appear to play a significant role. This behavior is readily explained by considering the fact that a magnetic field will tend to streamline the flow. When a slot exit is being used, there will not be a strong tendency for the flow to be disturbed in the lateral direction. This is in contrast to a point exit where the nature of the arrangement will promote flow nonuniformities.

4.5.2 Discussion

The results presented in this section show some interesting, and apparently unexpected, behavior.

When point inlet and point outlet are being used, there will be very strong short circuiting and by-passing, and quite a large fraction of the tundish volume will be inactive or "dead." Such a situation will arise in twin-roll casting and also in other novel continuous casting systems.

For these systems, the effect of the inlet and the outlet regions, which occupy a major fraction of the total tundish volume, are so significant that simple flow-control arrangements, such as a dam and weir or a narrow magnetic gate, will have only a small effect in improving the situation. However, the imposition of a magnetic field of say 1 kG or more across the whole tundish length can bring about spectacular improvement.

The case of a point inlet and a slot outlet, which would correspond to belt casters and to single-roll arrangements, is quite similar to the point inlet and point exit case, although perhaps a little less severe. Here again, there exists significant short circuiting and by-passing, and a strong recirculating flow pattern that can be controlled only by the imposition of, say, a 1 kG or higher field along the whole tundish. One may reasonably speculate that the imposition of a similar field over a significant length of the tundish would produce an intermediate result.

In summary, while in recent years a great deal of attention has been paid to the development of novel near-net-shape continuous casting systems, the tundishes associated with this technology have been rather less well studied. In this work, we have shown that flow phenomena in these shallow tundishes could be even more critical than those in conventional tundish arrangements, and have indicated the means for controlling the flows and for providing the highly desirable uniform velocities and long residence times.

As the initial problems with these near-net-shape casting systems are being solved, it is very likely that attention will turn to quality, including inclusion content. In this context, flow phenomena in tundishes could come to be of major importance.

4.6 Free Surface Phenomena and Wave Formation

In all previous work, we have explicitly assumed that the free surface of the melt in the tundish would be flat and undisturbed. This was thought to be a reasonable first approximation, as we wished to seek insight into the flow behavior in the bulk.

It is known, however, that there may be a dramatic increase in the inclusion count whenever ladle change takes place as seen in Fig. 4.48, taken from the work of van der Heiden et al. [8]. It is also possible that this feature is attributable to the flow disturbances that are inevitably created when the inflow or the outflow are being changed.

If we consider a vessel into which fluid is introduced at one end and from which fluid is being removed at the other, steady-state conditions will be established after some time. However, if we were either to stop the input or the exit stream suddenly, surface waves will form. This is a classical problem in hydrodynamics [15].

Problems of this type will arise in tundish operation. The modeling of the fully three-dimensional problem would be quite time consuming, but the following simplified, two-dimensional treatment should provide a helpful preliminary insight.

Let us consider a two-dimensional slice of the tundish, such as sketched in Fig. 4.49. Provision is made for feeding metal at one end through a submerged nozzle,

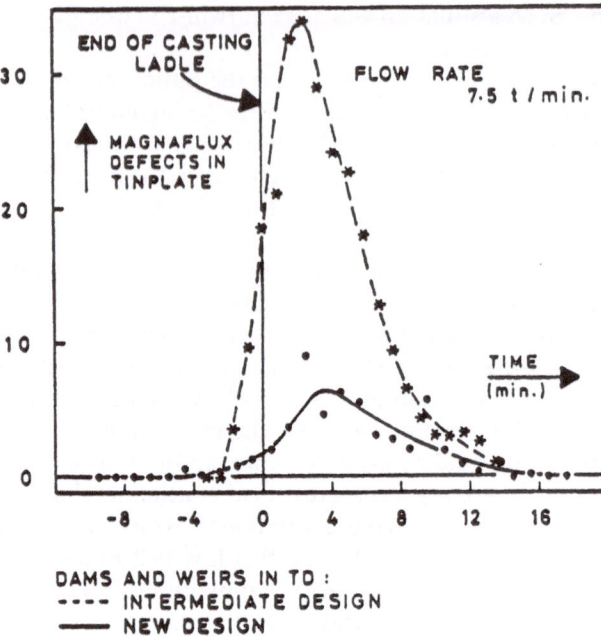

Fig. 4.48. Increase in magnaflux defects at ladle change (open teeming exluded); influence of dams and weirs in TD [8].

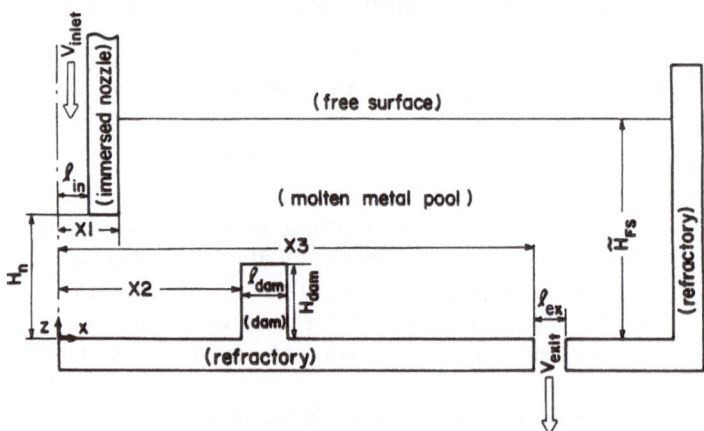

Fig. 4.49. A schematic sketch of tundish system in two dimensions.

Table 4.9. Operating parameters for tundish with allowance for wave formation at free surface

$X1 = 20$ cm	$l_{in} = 10$ cm	$H_{FS} = 57.72$ cm
$X2 = 60$ cm	$l_{dam} = 15$ cm	$H_{dam} = 0, 25$ cm
$X3 = 155$ cm	$l_{ex} = 10$ cm	$V_{inlet} = 0, -50, -75$ cm/s
$X4 = 200$ cm	$H_n = 40$ cm	$V_{exit} = 1, -50$ cm/s

and the molten metal stream is discharged through a bottom nozzle located "at the far end." The tundish is fitted with a dam as shown in the figure. The dimensions of the system and the typical operating parameters are summarized in Table 4.9.

Three types of tundishes were tested, namely,

1) deep pool ($H_{FS} = 72$ cm) with a dam ($H_{dam} = 25$ cm),
2) shallow pool ($H_{FS} = 57$ cm) with a dam ($H_{dam} = 25$ cm), and
3) deep pool ($H_{FS} = 72$ cm) without dam ($H_{dam} = 0$ cm).

In seeking to obtain insight into the behavior of the system, we shall consider the following somewhat idealized physical situations:

1) Initially stagnant melt in the tundish of a given height, and at time $= 0$ we start pouring in molten steel at a given flow rate, with no outflow of molten metal.
2) Steady-state operation, with a sudden change in the metal inflow rate, metal outflow kept constant.
3) The reladle situation. Initially no inflow, but steady-state outflow; then at time $= 0$ metal is poured into the tundish at the same rate as the outflow.

Within these three basic cases, we have considered the effect of the presence and the absence of a dam and different melt heights. In generating output, we were interested in examining both the flow behavior in the bulk and that of the free surface regions.

The governing equations are readily stated as the turbulent Navier–Stokes equations, written in two-dimensional Cartesian coordinate system. In allowing for turbulent behavior for the calculations here, we used a fixed value of the effective viscosity, which was 100 times larger than the atomic value.

Thus we have the following:

Continuity:

$$\frac{\partial U_x}{\partial x} + \frac{\partial U_z}{\partial z} = 0; \tag{4.17}$$

Equations of motion:

$$\frac{\partial U_x}{\partial t} + U_x \frac{\partial U_x}{\partial x} + U_z \frac{\partial U_x}{\partial z} = -\frac{1}{\rho}\frac{\partial p}{\partial x} + \frac{\mu}{\rho}\left(\frac{\partial^2 U_x}{\partial x^2} + \frac{\partial^2 U_x}{\partial z^2}\right) \tag{4.18}$$

and

$$\frac{\partial U_z}{\partial t} + U_x \frac{\partial U_z}{\partial x} + U_z \frac{\partial U_z}{\partial z} = -\frac{1}{\rho}\frac{\partial p}{\partial z} + \frac{\mu}{\rho}\left(\frac{\partial^2 U_z}{\partial x^2} + \frac{\partial^2 U_z}{\partial z^2}\right) + g. \tag{4.19}$$

Most of the boundary conditions are readily stated as

$$U_x = 0, \quad U_z = V_{inlet} \qquad at \quad 0 \le x \le l_{in}, z = H;$$

$$U_x = 0, \quad \frac{\partial U_z}{\partial x} = 0 \qquad at \quad x = 0, 0 \le z \le H;$$

$$U_x = 0, \quad p = 0 \qquad at \quad X3 < x < X3 + l_{ex}, z = 0;$$

$$U_x = 0, \quad U_z = 0 \qquad at \quad \text{all solid surfaces;}$$

that is, the velocities were specified at the inlet, zero velocities were specified at all solid surfaces, and axial symmetry was observed. The boundary conditions on the velocity at the free surface are detailed in Ref. 16.

The key factor in this formulation is the modeling of the free surface. The behavior there was represented by the SOLA-VOF technique, which is essentially a modified "marker and cell" method. The principle upon which the technique is based is discussed in detail in Ref. 17 and 18. A brief description would be to state that the region near the free surface is divided into cells. Upon establishing a mass balance over these cells, the upper boundary may expand or contract in an appropriate manner, while also satisfying a force balance.

Moreover, in order to estimate the shape of the free surface, the fractional volume of fluid (VOF) was introduced as expressed below:

$$\frac{\partial f}{\partial t} + U_x \frac{\partial f}{\partial x} + U_z \frac{\partial f}{\partial z} = 0. \tag{4.20}$$

This equation states that f moves with fluid. The average value of f in a cell is equal to the fractional volume of the cell occupied by the fluid. The "donor–acceptor" method was used to solve Eq. (4.20).

In generating the solution, 40 grid points were used in the horizontal direction and 20 in the vertical direction. At this level of grid points, the results were no longer grid sensitive. The typical time step used was 0.01 s. The SOLA-VOF technique was used to generate the solution, which took about 5 h of CPU time to represent about 30 s of real-time operation on a Micro VAX II digital computer.

4.6.1 Computed Results

In the following, we shall present a selection of the computed results by considering the three basic cases described above.

Figures 4.50 and 4.51 show the behavior of the system, which is initially stagnant. Then, at time = 0, metal is poured in at the entrance nozzle, but no metal is being discharged at the "other end." Figure 4.50 shows the horizontal velocity component as a function of the vertical position, and time, halfway between the entrance and the exit. The four cases A, B, C, and A' are explained in the captions. The oscillatory motion that is being set up is readily apparent on inspection of Fig. 4.50. It is seen, futhermore, that the disturbances seem to manifest themselves throughout the whole tundish depth.

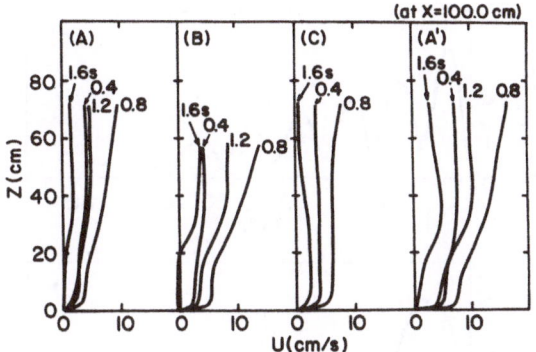

Fig. 4.50. Changes in the horizontal velocity at $x = 100$ cm (initially stagnant state): (A) deep pool (72 cm) with a dam, $V_{inlet} = -50$ cm/s, $V_{exit} = 0$; (B) shallow pool (57 cm) with a dam, $V_{inlet} = -50$ cm/s, $V_{exit} = 0$; (C) deep pool (72 cm) without a dam, $V_{inlet} = -50$ cm/s, $V_{exit} = 0$; (A') deep pool (72 cm) with a dam, $V_{inlet} = V_{exit} = -50$ cm/s.

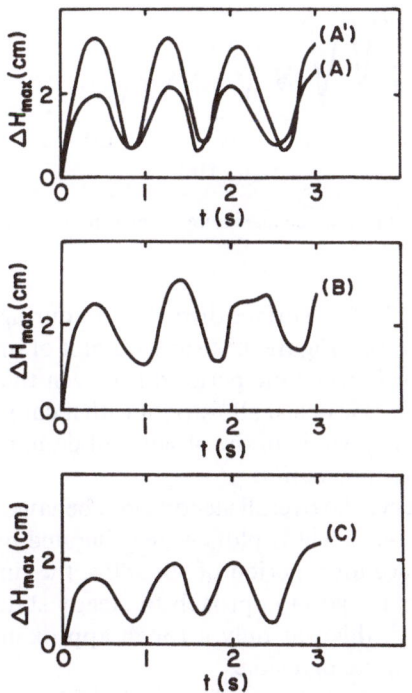

Fig. 4.51. Changes in the maximum disturbance of the free surface.

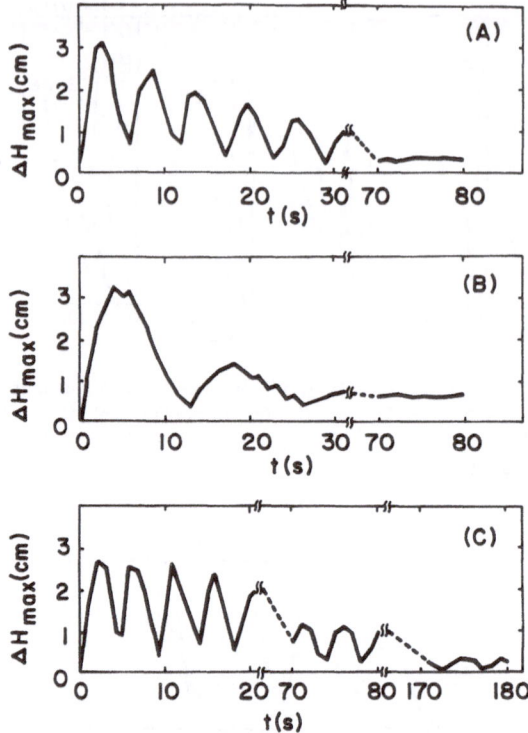

Fig. 4.52. Damping of the maximum disturbance of the free surface.

Figure 4.51 illustrates the free surface disturbances during the first 3 s, which are seen to be quite considerable. Figure 4.52 shows a plot of the maximum amplitude as a function of time, for longer time periods; it is seen that the disturbances will decay quite rapidly in the case of a shallow pool with a dam. The decay is less rapid for a deep pool with a dam, while in the absence of dam, the oscillations are seen to persist for quite a long time period.

It is of interest to observe the overall steady-state behavior of these systems. This is depicted in Figs. 4.53–4.55. These plots were generated by running the program after an initial disturbance for a period of 80–180 s, the time needed to approach steady state. It is felt that a good approach to steady state has been achieved in cases A and B, but for C, this was only a rough approximation. The streamlines seen in these plots seem quite plausible.

The dynamic behavior depicted in Figs. 4.50–4.52 was for a highly idealized situation, with sudden flow imposed on a system initially at rest. This would be expected to give the maximum disturbance.

Figures 4.56–4.58 illustrate a more realistic situation, when a sudden increase in the flow rate is imposed on a system for an initial condition of steady throughflow. It should be remarked that these "steady" conditions corresponded to running the system, starting from rest, for an extended period corresponding to plots shown in

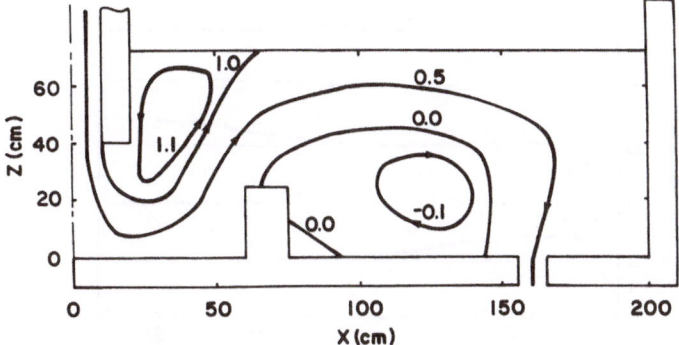

Fig. 4.53. The computed normalized streamlines under steady state for case A.

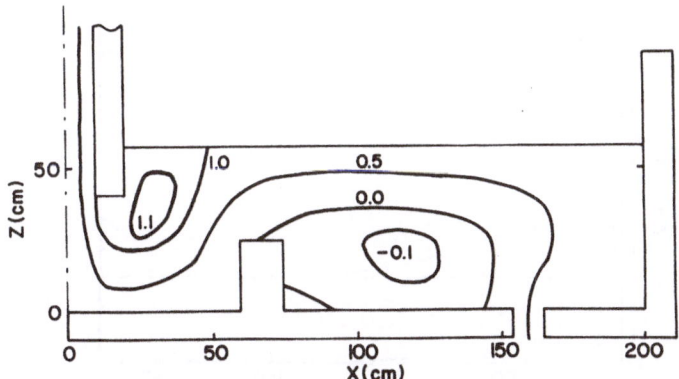

Fig. 4.54. The computed normalized streamlines under steady state for case B.

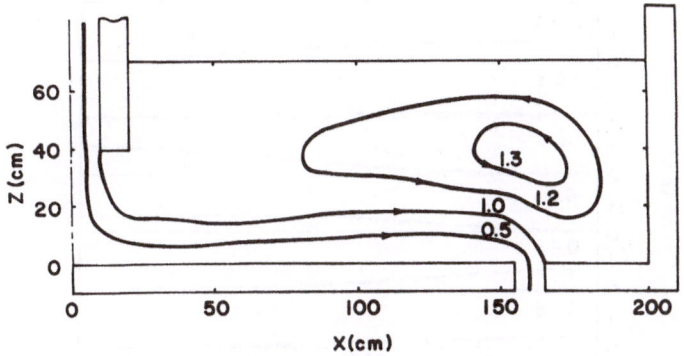

Fig. 4.55. The computed normalized streamlines under steady state for case C.

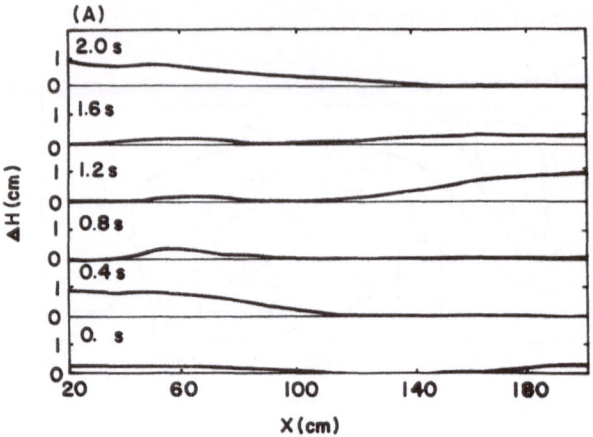

Fig. 4.56. The behavior of the free surface for case A after V_{inlet} is changed from -50 to 75 cm/s.

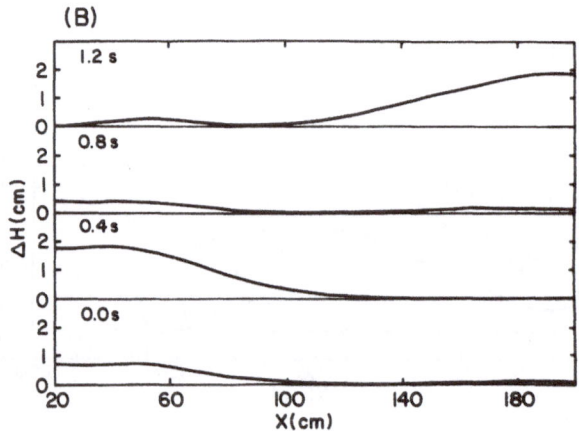

Fig. 4.57. The behavior of the free surface for case B after V_{inlet} is changed from -50 to 75 cm/s.

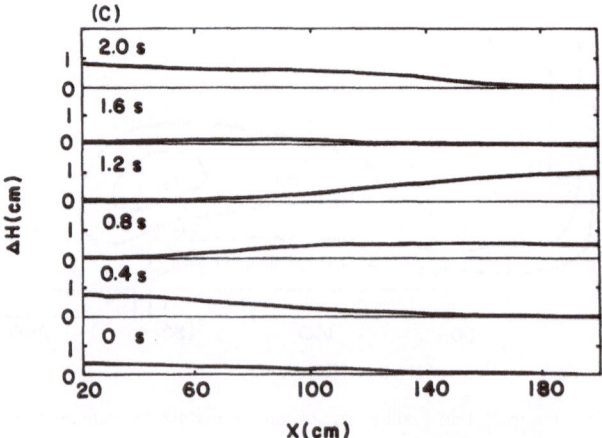

Fig. 4.58. The behavior of the free surface for case C after V_{inlet} is changed from -50 to 75 cm/s.

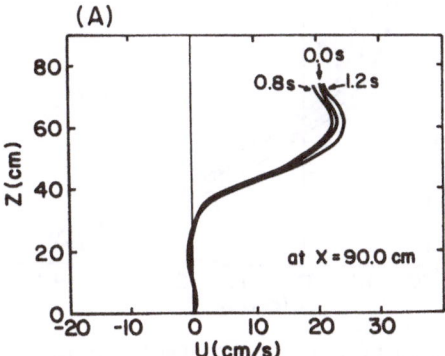

Fig. 4.59. Changes in the horizontal velocity at $x = 90$ cm after V_{inlet} is changed from -50 to 75 cm/s for case A.

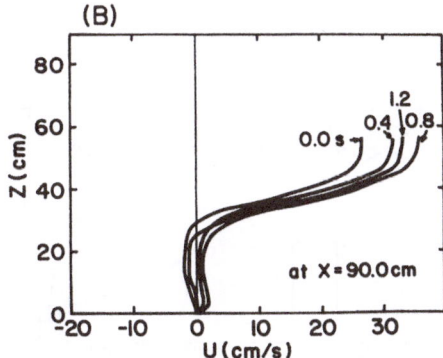

Fig. 4.60. Changes in the horizontal velocity at $x = 90$ cm after V_{inlet} is changed from -50 to 75 cm/s for case B.

Figs. 4.53–4.55. As seen earlier, this was quite a good approximation for cases A and B, but rather less satisfactory for case C.

Figures 4.56–4.58 show the free surface behavior of systems A, B, and C corresponding to a deep pool with a dam, a shallow pool with a dam, and a tundish with no dam. The characteristic standing wave motion is readily apparent on these plots. As expected, we have higher amplitudes in the shallow pool.

Figures 4.59–4.61 show the vertical variations in the horizontal velocity during a short initial time period for cases A, B, and C. It is seen that for A, the oscillations get readily damped out as we progress downward from the free surface, but that the disturbances do propagate well into the tundish depth for the other two cases.

Finally, Figs. 4.62–4.65 depict a situation where the steady throughflow of material is disturbed by shutting off the inlet while maintaining the outflow. These cases correspond to the realistic situation of a ladle change.

Figure 4.62 shows the steady-state streamline pattern, which seems quite plausible. Figure 4.63 shows a very marked oscillatory flow behavior in the hori-

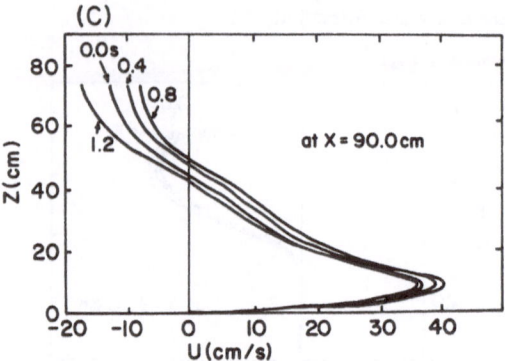

Fig. 4.61. Changes in the horizontal velocity at $x = 90$ after V_{inlet} is changed from -50 to 75 cm/s for case C.

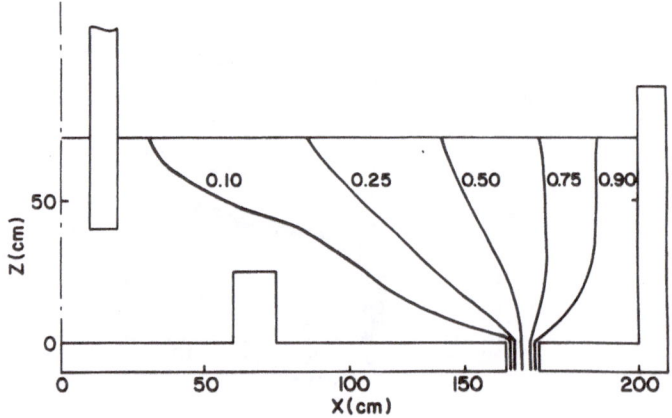

Fig. 4.62. The computed normalized streamlines under steady state for case A but with $V_{exit} = -50$ cm/s.

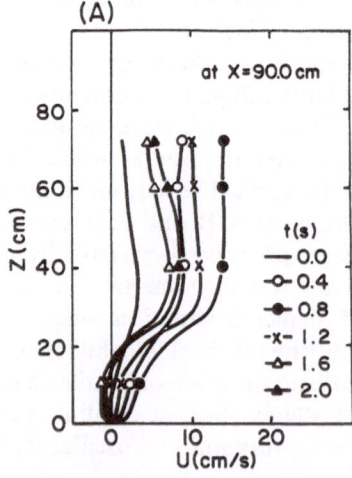

Fig. 4.63. Changes in the horizontal velocity at $x = 90$ cm after V_{inlet} is changed from 0 to -50 cm/s.

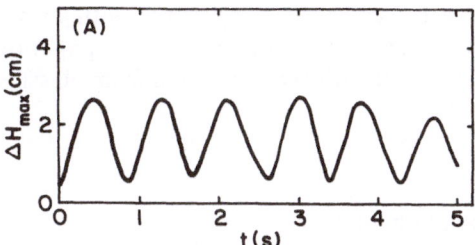

Fig. 4.64. Changes in the maximum disturbance of the free surface after V_{inlet} is changed from 0 to -50 cm/s for case A.

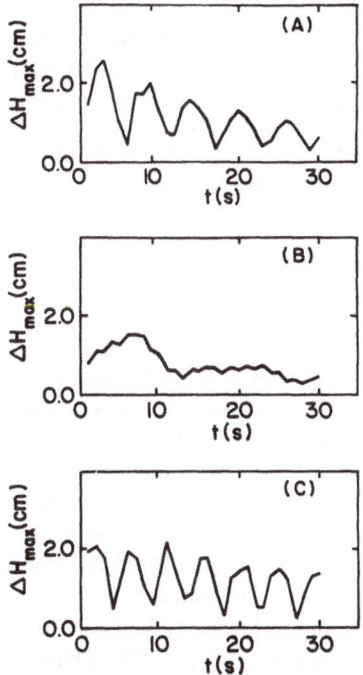

Fig. 4.65. The damping of maximum disturbance of the free surface after V_{inlet} is changed from 0 to -50 cm/s for case A, B, and C.

zontal velocities that seems to extend to the bottom of the tundish. Figure 4.64 shows the very marked free surface oscillations that are being produced, and their decay for all three cases is illustrated in Fig. 4.65.

4.6.2 Discussion

The computed results have shown that significant surface oscillations will occur whenever there is a change in the flow conditions, that is, when the inflow is started or increased. These oscillations produce surface waves with a typical amplitude in the 1–2 cm range.

These waves appear to be typical standing waves, with a wavelength of the same order as that of the length of the vessel; the speed of propagation is about 250 cm/s, which aggrees reasonably well with the prediction from classical wave theory:

$$v_w = (gh)^{1/2}, \tag{4.21}$$

where

$v_w \equiv$ speed of propagation, and

$h \equiv$ pool depth.

The results have shown that the disturbances appear to be the most severe for the shallower pools and in the absence of dams; this behavior is in accord with expectations.

From the standpoint of practical operations, the maximum disturbances are likely to occur when flow is shut off or when pouring commences.

The following are the principal conclusions that may be drawn from this work.

1) Some form of wave motion will inevitably occur in tundish operation whenever the metal inflow is increaded or shut off, that is, during ladle change. Typically, the length of these waves may be of the same order of magnitude as the length of the vessel, with a speed of propagation of about 250 cm/s.
2) The use of dams (or baffles) would tend to damp out the wave motion, and by the same token, the disturbances are likely to be less severe for the deeper pools.
3) The actual time period for which these waves would persist is relatively short, ranging from a few tens of seconds to a few hundreds of seconds. This time period is short compared with a typical ladle-emptying cycle, which is of the order of tens of minutes for most practical systems.
4) Wave motion is likely to have an adverse effect on the quality of the continuously cast products, because slag may be entrained and the instabilities, if they persist through the melt depth, could cause problems with the solidified structures. The use of suitably placed dams or weirs would help to minimize wave formation and aid in the more rapid damping of the waves.
5) It is an established fact that the impurity levels in the cast product tend to rise during ladle change. This behavior is possibly attributable to slag entrainment and the general flow instabilities associated with emptying the last parts of the molten metal charge from a ladle. However, a significant contribution from the associated wave formation may also be a strong possibility.
6) It is felt that this work represents a first attempt at studying these phenomena, and a great deal of further work, including experimental studies, would be fully justified.

References

1. B.E. Launder and D.B. Spalding, Computer Methods Appl. Mech. Eng., 3, 269(1974).
2. B.E. Launder and D.B. Spalding, Mathematical Models of Turbulence Academic Press, London (1972).

3. K.H. Tacke and J.C. Ludwig, Steel Research, *58*, 262–270 (1987).

4. J. Szekely and N.J. Themelis, Rate Phenomena in Process Metallurgy, Wiley-Interscience, New York (1971).

5. D.B. Spalding, Mathematics and Computers in Simulation, *XIII*, 267–276 (1981).

6. S.V. Patankar and D.B. Spalding, Int. J. Heat and Mass Transfer, *15*, 1786 (1982).

7. J. Szekely, O.J. Ilegbusi, and N. El-Kaddah, "The Mathematical Modelling of Complex Fluid Flow Phenomena in Tundishes," PCH PhysicoChemical Hydrodynamics, *9*, 3–4, 453–472 (1987).

8. A. van der Heiden, P.W. van Hasselt, W.A. de Jong, and F. Blaas, "Inclusion Control for Continuously Cast Products," in Proceedings of the 5th International Iron and Steel Conference, Washington, DC, Iron and Steel Society, pp. 755–760 (1986).

9. O.J.Ilegbusi, J. Szekely, R. Boom, A. van der Heiden, and J. Klootwijk, "Physical and Mathematical Modelling of Fluid Flow and Tracer Dispersion in a Large Tundish and a Comparison with Measurements in Hoogoven's System," Proceedings, W.O. Philbrook Memorial Symposium, Toronto, Canada, April 17–20, 1988.

10. T. Saeki, O. Tsubakihara, A. Kusano, K. Umezawa, and I. Suzuki, "The Roles of Tundishes in Continuous Casting of Steel," Proceedings of the Japan Iron and Steel Society (1987).

11. ASEA Industrial Systems Brochure (1987).

12. J.A. Shercliff, A Textbook of Magnetohydrodynamics, Pergamon Press, Oxford (1966).

13. J.P. Birat, "Manufacture of Flat Products for the 21st Century," Proceedings of the Conference on Restructuring Steel for the Nineties, Institute of Metals, London, p. 140 (1986).

14. J.D. Naumann and D.B. Love, "Controlling the Steel Shape of Cast Stainless Steel Strip," Presented at the International Symposium on Near-Net Shape of Casting Strip, 89th Annual Meeting of CIM, Toronto (May 1987).

15. H. Lamb, Hydrodynamics, Dover Publications, New York (1945).

16. B.D. Nichols, C.W. Hirt, and R.S. Hotchkiss, Los Alamos Scientific Laboratory Report LA-8355 (August 1980).

17. F.H. Harlow and J.E. Welch, Phys. Fluids, *8*, 2182 (1965).

18. F.H. Harlow, J.E. Welch, J.P. Shannon, and B.J. Daly, Los Alamos Scientific Laboratory Report LA-3425 (1966).

19. O.J. Ilegbusi and J. Szekely, Fluid Flow Phenomena and Tracer Dispersion in Shallow Tundishes, Verlag Stahleisen mbH, Dusseldorf, West Germany (1988).

5 Concluding Remarks

In this monograph, we sought to introduce the reader to the principles that govern the mathematical and physical modeling of tundish operations.

It is an established fact that tundishes may have an enormous impact on the performance of a given steel plant. In most cases, we may have literally several million dollars worth of steel passing through a given tundish system annually. For a properly designed tundish, steel of the required inclusion count and specified superheat level would be supplied at a steady, smooth rate to the caster. If these conditions are not met, serious production problems may result. This may be especially the case for the higher, more demanding steel grades.

Notwithstanding these facts, in many cases very little science or engineering is being used in tundish design and operation, and possible changes are contemplated only when an already finalized construction leaves very little flexibility.

As is the case with most metals and materials processing systems, the operation of tundishes involves a complex array of chemical and physical phenomena, including (transient) turbulent fluid flow, heat and mass transfer, and possibly magneto-hydrodynamics.

In this monograph, we sought to describe some of the tools that are available for the study of tundish behavior and to discuss the insights that may be obtained from the work that has been done previously.

In the following, we shall summarize the major findings that have been presented in the text and will conclude with a brief discussion as to what may constitute an optimal tundish design and operation strategy.

5.1 Main Findings

The main findings of the work may be summarized as follows:

1) For most tundish systems, the flow is highly turbulent in the inlet and in the exit regions and will be in a transitional state in the bulk of the tundish.

2) It is an inherent feature of most tundish geometries that dead or inactive zones may make up at least one-third and at times as much as three-quarters of the total volume. Designs where the inlet and the exit ports are close together should be always avoided. The volumes between the exit port and the "far wall" will be inactive and hence quite useless.

   In general, long, relatively narrow, and too deep tundish arrangements should be preferred to short squat geometries. By the same token, larger tundishes would be preferred to smaller ones, but there has to be an optimal size, because greater heat losses will occur in larger tundishes. Furthermore, if sequence casting is not being employed, there will be a loss in quality as the tundish is being emptied.

3) Some form of flow control with dams and weirs is almost always desirable, because a tundish in the absence of flow-control arrangements will be prone to short circuiting and by-passing. However, too many dams will also create an excessive dead zone region.

4) Tundish heaters and magnetic flow control are recent developments, which should be given consideration, because they may help in significantly upgrading the tundish performance. The optimal deployment of these devices needs an accurate knowledge of the transport phenomena in the tundish.

5) All the previously mentioned considerations will apply to the shallow tundishes used in novel continuous casting operations, and here the tundish design may be even more critical. We should also stress that for the shallow tundishes, the flow field tends to be dominated by the inlet and the outlet arrangements and so the use of conventional flow-control devices may not be adequate.

5.2 Strategy for Optimal Tundish Design

Having made the preceeding remarks, let us conclude with a brief discussion of what may constitute an optimal tundish design strategy.

1) It is to be recognized that the tundish is a critical part of the steel processing operation, so that great care has to be taken in allowing for an optimal tundish geometry right at the outset. Often the continuous casters and the steel delivery system are fixed and the tundish has to "fit in between"; this is putting the cart before the horse.

2) In general, one should decide on the tundish shape, and capacity, through the intelligent use of mathematical and physical modeling and have this as an input into the layout considerations.

3) The very major impact of tundish performance on product quality should readily justify a reasonable resource allocation to having an optimal design identified. In doing this work, mathematical and physical modeling should be used in complementary fashion.

4) In physical modeling, water is an ideal medium and the similarity criteria mandate that the model should be of the same scale as the prototype.

5) Mathematical modeling is now reaching a mature state, and we can calculate with confidence the velocity fields and temperature profiles, and can estimate reasonably well how inclusions would behave in tundish systems.

6) Ideally, this mathematical and physical modeling work should be done as part of the design process. However, the modeling approach should be quite useful also in optimizing existing systems. In particular, flow control devices should be retrofitted if these were not installed originally.

7) There is now an excellent, growing science base for evolving optimal tundish design and operation, and it is incumbent on the technical community to make good use of this information.